SYMPOSIA OF THE
BRITISH SOCIETY FOR PARASITOLOGY
VOLUME 10

FUNCTIONAL ASPECTS OF PARASITE SURFACES

EDITED BY

ANGELA E. R. TAYLOR

Lister Institute of Preventive Medicine
Chelsea Bridge Road, London, SW1W 8RH

AND

R. MULLER

London School of Hygiene and Tropical Medicine
Keppel Street, London, WC1

BLACKWELL SCIENTIFIC PUBLICATIONS

OXFORD LONDON EDINBURGH MELBOURNE

© 1972 by Blackwell Scientific Publications
5 Alfred Street, Oxford, England
3 Nottingham Street, London W1, England
9 Forrest Road, Edinburgh, Scotland
P.O. Box 9, North Balwyn, Victoria, Australia

ISBN 0 632 08920 2

First published 1972

Distributed in the U.S.A. by
F. A. Davis Company, 1915 Arch Street,
Philadelphia, Pennsylvania

Printed and bound in Great Britain by
Burgess & Son (Abingdon) Limited
Abingdon, Berkshire

CONTENTS

PREFACE

The exceptionally large attendance at this year's autumn meeting of the British Society for Parasitology (held at the Zoological Society, London, on 22 October, 1971) provided ample evidence of the popularity of the subject, the functional aspects of cell surfaces. We were fortunate in having Professor A. S. G. Curtis and Dr A. C. Allison to discuss the subject from the non-parasitologist's point of view, and the lively discussions following all the papers indicated once more that inter-disciplinary meetings are of great benefit to members of the Society.

I would like to thank Professor A. S. G. Curtis and Dr B. A. Newton for so kindly chairing the morning and afternoon sessions respectively and also all the speakers and discussants for providing such a stimulating meeting.

ANGELA E. R. TAYLOR

Lister Institute of Preventive Medicine,
Chelsea Bridge Road,
London SW1W 8RH

ADHESIVE INTERACTIONS BETWEEN ORGANISMS

A. S. G. CURTIS

Department of Cell Biology,
University of Glasgow

One of the fundamental factors which may determine the establishment of communities of living organisms is the ability or inability of the animals or plants to form adhesions with one another. There are two ecological niches in which this problem is particularly important—the first is underwater, in particular on rocky substrates, where colonial organisms can abut on or overlap each other. For example, settlement may be determined by the presence or lack of the ability to form an adhesion with a given rocky substrate. Further settlement on top of the first organisms will be controlled in a similar manner as will the extent to which organisms can overlap each other or even form lateral adhesions. These factors will also determine the mechanical resistance of the community to erosion by currents. The second ecological niche is that of the parasite. Failure to develop an adhesion with the host may prevent the establishment of a parasite. Conversely, a strong adhesion may lead to encystment or even phagocytosis of a parasite. Loss of adhesion between host and parasite will often result in elimination of the latter. Thus a complex series of changes in adhesion may be necessary during the life cycle of a parasite. The purpose of this short review is to outline the possible mechanisms by which these interactions may be determined and to suggest possible means of developing control of parasite-host relationships.

Parasite-host interactions are often highly specific. It might be suspected that this specificity could reside in the adhesive interactions of the organisms. If cells can adhere specifically to one another this mechanism might be of importance in the consideration of parasitism. Specific adhesion has been suggested as being a widespread phenomenon (Wilson, 1907; Moscona, 1962; Burnet, 1971). For this reason the latter part of this review is devoted to a consideration of the phenomenon.

FACTORS CONTROLLING ADHESION

The development of an adhesion between two bodies requires (a) the existence of a mechanism for bringing the particles into a contact sufficiently close for (b)

1

forces of attraction to operate between them. In addition (c) the area of adhesion must be large enough to provide sufficient adhesive strength to resist the usual forces in the environment which might break the bond. This may mean, for example, that two very rigid bodies which can form very strong point to point adhesions are unable to establish a sufficient area of contact to form an effective adhesion; while two very deformable cells of rather low adhesiveness can form such an effective adhesion because they can extend the area of contact.

Collision between two cells or particles will be brought about by (i) active motile movements of one or both of the cells or organisms involved, (ii) shear forces due to velocity gradients in the surrounding medium (see Curtis, 1969, 1970a), (iii) gravity and (iv) Brownian motion. Only very small particles of the size of bacteria and below are likely to be sufficiently influenced by Brownian motion for an appreciable number of collisions to be produced by this mechanism. All factors inducing collision tend to propel the particles apart after collision unless the impact is in a normal direction. Thus an adhesion must form rapidly so that the subsequent tendency of the particles to pull apart is overcome. If particles are brought together by a great velocity difference the mechanical forces tending to aid adhesion are great but the lifetime of approach is short, to be followed by equally great mechanical forces tending to pull the two particles apart. If the velocity difference is small the lifetime of the interaction is long but the mechanical forces at first aiding and then tending to oppose adhesion are small. Thus it can be seen that the factors controlling a collision, which may bring the particles sufficiently close for a long enough time for an adhesion to form, are complex. In addition the medium between the particles must drain away into the bulk region of the medium as the particles approach. Rapid movement of the medium out of the narrow gap between two approaching particles requires appreciable energy input. A complete solution to this problem for laminar shear flow conditions has been produced by Curtis and Hocking (1970). This has permitted the development of a means of evaluating the collision probability, that is the probability that two particles will stick together on collision, in terms of the adhesive energy of interaction of two cells. Collision probability measurements provide a powerful method of measuring the adhesiveness of particles. The practical aspects of the technique are described by Curtis (1969). Analogous treatments could be developed for turbulent flow and Brownian motion induced collision systems but so far only a special case of the latter has been examined (Fuchs, 1934), which might be appropriate for the investigation of the collisions of aerosol particles of bacteria with one another.

The separation of two cells or organisms requires the breaking of an adhesion by the application of sufficient mechanical force. Brownian motion will only be effective if the adhesive energy of the contact between the cells is very low. In practice the breaking of an adhesion is a complex phenomenon for investigation because deformation of the cells is likely to occur. In consequence measurement of cell adhesion from the magnitude of the force required to separate two cells is a somewhat imprecise technique.

The formation of an adhesion depends upon the natures of the surfaces involved. Many parasites carry an external cuticle so that the adhesive interaction of parasite and host involves at most only one cell surface, that of the host. However, in other instances the cell surfaces of both parasite and host are involved in the adhesion. For this reason, and because the chemistry of cell surfaces does not differ essentially from that of many other surfaces, it is appropriate to consider the role of cell surface chemistry in adhesion at this point.

The nature of the cell surface is still unclear though much work has been carried out in recent years (see reviews by Hendler, 1971; Zahler, 1969; Curtis, 1972a). One of the most generally accepted models is that of a lipid bilayer structure penetrated by protein molecules which may traverse the full thickness of the lipid (Fig. 1).

Electron microscopic observation of many cell types has shown the presence of a faint line of staining outside the plasmalemma; this is the surface coat or glycocalyx. This structure should not be confused with the basement membrane (Goel and Jurand, 1971). Two extreme views can be held concerning the existence of the glycocalyx. The first is that it really exists separate from the plasmalemma in life and that it coats the cell to a depth of circa 200 Å. The second view is that the glycocalyx is largely an artefact of fixation. If the first view is correct cells should make contact when their plasmalemmae are separated by the thickness of two surface coats. Biochemical work bears on this problem because it shows that much of the glycosidic material associated with the cell surface is carried either by the glycolipids, whose glycosidic portions are short, or by glycoproteins where the chain length may be relatively short (Cook *et al.*, 1960) and which probably have the greater part of their structure running through the membrane from one side to the other (Bretscher, 1971). These glycosidic regions will be hydrophilic because of the large number of hydroxyl groups they carry. Many of the antigenic properties of the cell will be determined by these groupings (Pardoe and Uhlenbruck, 1971). Thus it seems likely that such cell surface associated glycoprotein and glycolipid as can be detected by present biochemical techniques extends only some 15–20 Å from the lipid at most. The glycoproteins appear to be so closely associated with the cell surface lipid that it is exceedingly difficult to regard them as a separate structure. If all surface associated glycosidic material should turn out to have these two arrangements the term glycocalyx will be redundant. Electrophoretic evidence (see discussion by Curtis, 1967) suggests that the charged groups of the cell surface lie in a fairly narrow zone close to the lipid phase of the plasmalemma. A large proportion of these charged groups are N-acetyl-neuraminic acid or other neuraminic acids in many cells. This sugar is often placed terminally on glycosidic chains.

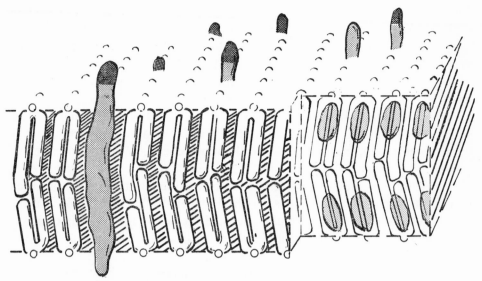

Fig. 1. Cell surface structure. This diagram shows a small portion of the surface in perspective view with the outer side of the surface on the top and a section through the thickness of the membrane in front. This structure is largely based on recent work reviewed by Curtis (1972a). Phosphatidyl and sphingomyelin components are shown unshaded and their hydrocarbon chains appear as long cylinders. Note that their long axes are slightly inclined and that components on either side of the membrane tend to interdigitate. Their polar head groups are shown as small circles. Proteins (middle degree of shading) run through the thickness of the membrane and the majority bear glycosidic terminations (dark shading) terminally on the outer side of the membrane. Some of the glycosidic terminations are attached to lipids as glycolipids, e.g. ceramide. The cholesterol molecules (lightest dotted shading) lie in between the rows of phosphatidyl and sphingolipid and can be seen at the right-hand end of the section. These glycosidic terminations and the occasional naked protein extend up to 20Å from the surface and may be equivalent to the glycocalyx. It should be remembered that the surface is in active molecular motion at physiological temperatures and that the diagram shown here has greater order than would normally be found. However, examination of the surface will show that some attempt has been made to show some of this motion because waves of dislocation of the headgroups can just be seen.

Thus electrophoretic evidence does not support the concept of a glycocalyx. If, however, the concept of an extended glycocalyx is correct then we would expect any gap of up to circa 400 Å between two plasmalemmae to be filled with this carbohydrate material, which might act to stick cells together. Consequently, the viscosity of the material in the gap between cells would be high. Brightman (1965) however found that ferritin would permeate rapidly down such gaps in a way which suggests that there was no carbohydrate gel in the gap between plasmalemmae. However, if the cells were fixed before permeation was attempted, ferritin would not permeate, which suggests that a gel-like material may be released into the gap during fixation. Loss of and changes in the surface coat have been observed by a number of investigators (e.g. Vickerman, 1969). This does not prove the existence of a surface coat because it may be no more than a reflection of changes in surface permeability after fixation or alterations in the ease with which surface glycoproteins change their conformation on fixation.

It seems possible that the glycocalyx is nothing more than the glycoprotein and glycolipid part of the cell surface swollen and partially discharged from the plasmalemma after fixation for electron microscopy. If such is the case it might be that those cells which appear to lose their glycocalyces (see Clegg, p. 24) have in fact developed plasmalemmae which contain very small amounts of glycosidic components or have glycosidic components that are resistant to disruption on fixation.

Thus it appears that the cell surface carries a number of anionic charges in a comparatively narrow zone outside the lipid barrier. The lipid region has a low dielectric constant and a high electrical resistance. These anionic charges are partly due to the presence of neuraminic acids and may be partly contributed by species such as chloride adsorbed from the medium. Other surfaces will probably be similar in that they will have a hydrophobic core and will carry a certain surface charge, even though the precise details of their surface chemistry will obviously be different. The charges on the surfaces give rise to electrostatic forces of repulsion between like surfaces. These forces may play some part in controlling the interaction of particles.

Adhesive interactions can be controlled either by altering the forces of attraction or the forces of repulsion between particles. One set of questions of importance are centred round the nature of the components of the cell or other surfaces that are of importance in determining the adhesive interactions of cells and organisms, and whether these components act by altering the forces of attraction or those of repulsion. The answers to these problems will appear with more clarity when the mechanism of adhesion has been considered.

There are basically three types of mechanism of adhesion, (i) adhesion by sensitisation (bridging or cementing), (ii) adhesion by molecular contact and (iii) adhesion in the secondary minimum (of the potential energy diagram for interaction). Adhesion by bridging requires that the surfaces be bridged by molecules that are bi- or polyfunctional so that one end of the bridging agent binds to one surface and the other to the opposite surface. Antibody-induced agglutination is an example of this type of adhesion. Typical of this type of adhesion are (a) a gap is found between the particles whose breadth corresponds to the mean distance between functional groups on the bridging agent, (b) the strength of the adhesion will be determined by the relative proportions of bridging agent and binding sites. If the agent is bivalent maximal adhesiveness will be found when there is sufficient agent to bind with half the binding sites on every surface: the addition of further binding agent to a system in the process of forming adhesions will weaken adhesion, (c) adhesions will tend to be very strong unless the density of bonds between surfaces is at a very low value. (d) It will produce a type of bonding that is very sensitive to specific changes in chemical conditions. Very specific forms of adhesion might result.

Adhesion by molecular contact may seem a slightly misleading title because clearly bridging adhesions have molecular contacts. The use of this term by colloid chemists has come to refer to those adhesions where a surface, which is a narrow zone of marked transition in physical properties such as dielectric constant, abuts directly on another. Forces arising at least in part from the transition across the surface hold the surfaces together. Such adhesions are characterised by the fact that there is a very small gap between the surfaces and by their extreme strength and irreversibility.

The secondary minimum type of adhesion is characterised by the presence of a gap varying between 60 Å upwards to 300 Å (for the ionic and other conditions in which most cells live) between the surfaces. This gap is filled with the medium that surrounds the particles. To those who are unfamiliar with colloid chemistry it may seem surprising that such adhesions could exist. The reason for their existence is explained by the Derjaguin-Landau-Verwey-Overbeek (DLVO) theory of colloid adhesion (see Visser, 1968; Curtis, 1962). Adhesion, according to this theory, is controlled by the interaction of electrostatic forces of repulsion and the attraction arising from the London–van der Waals dispersion force. This latter force arises from the electronic and molecular vibrations of individual atoms and molecules. Recent papers by Napper (1967) and by Ninham and Parsegian (1970) describe this force in detail. Visser (1968) reviews recent theoretical and experimental work on this force.

Since the interaction energies arising from the electrostatic repulsion and the dispersion attraction decline with increasing distance between the surfaces in differing ways, it is possible to have interaction curves (for the sum of the two separate systems) of the type shown in Fig. 2. It can be seen that this diagram indicates that adhesion will occur for two situations. The first, primary minimum adhesion, is found when there is no gap between the surfaces. This corresponds to adhesion by molecular contact. The second position for adhesion is the secondary minimum in the potential energy diagram; which occurs with a gap between the surfaces when the electrostatic energy of repulsion at this separation is less than the London energy of attraction. A potential energy barrier due to electrostatic energy of repulsion may be found, preventing a secondary minimum type of adhesion becoming a primary one (equivalent to adhesion by molecular contact). If the surfaces are uncharged this barrier will be absent. If the surfaces are very highly charged the forces of repulsion will increase and the depth of the secondary minimum will become so small that it will not provide sufficient energy for any effective adhesion.

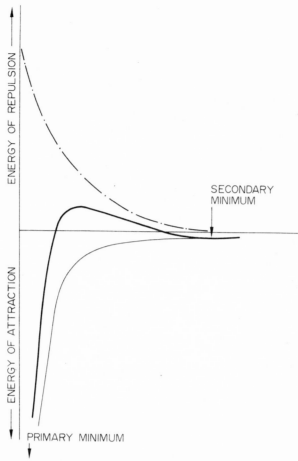

FIG. 2. Potential energy diagram for the interaction of two surfaces according to the DLVO theory. Ordinate, energy of interaction, repulsion positive, attraction negative. Abscissa distance between the surfaces. Electrostatic energy of repulsion———·———·——— Van der Waals energy of attraction, ———·———; total energy of attraction, ————————. This diagram shows the origin of the primary and secondary minima.

Consequently changes in the electrostatic forces due to increase or decrease in surface charge, or to change in counter-ion valency or concentration in the surrounding medium will affect adhesion. Changes in the di-electric constant of the surrounding medium or of the cell surfaces will affect both the forces of attraction and those of repulsion.

Adhesions of this type will have the following characteristics: (i) the presence of a gap between the plasmalemmae. The dimensions of the gap will be sensitive to any treatment affecting either the electrostatic forces or those of attraction, (ii) the adhesions will not be specific, (iii) they will be of low energy and consequently easily broken.

The evidence for biological systems in favour of each type of adhesion has been reviewed in recent years by Curtis (1967), Weiss (1967) and more recently by Curtis (1972b). It arises from the following main fields of investigation: (a) examination of the morphology of contacts between cells and of cells with other surfaces, mainly by electron microscopy, (b) measurement of the energy of cell adhesion, (c) study of the conditions under which adhesion can be weakened by various chemical treatments and (d) examination of the conditions under which adhesions form. Specificity of adhesion, if it occurs, would provide powerful evidence against the secondary minimum interpretation and would tend to support the bridging hypothesis. However, the question of the existence of specific adhesion will be discussed later in this review.

At present there is no unequivocal evidence in favour of, or against, any of the three main theories. The morphology of cell contacts (see reviews by Farquhar and Palade, 1963; Kelly and Luft, 1966) tends to support the idea that there are two types of adhesion, the first in which the lipid-zones of the plasmalemmae come into molecular contact and the second in which a gap is found between the plasmalemmae. The former type is of very great adhesive energy while the gap adhesions are weak and easily broken. Curtis (1962, 1967 and in press) has argued that the gap adhesions represent secondary minimum interactions, and that the zonulae occludentes, in which the two lipid surfaces come into molecular contact, represent primary minimum situations. The dimensions of the gap agree well with the predictions of the DLVO theory. If the glycocalyx has a definite existence as a thick layer separate from the cell surface it might be reasonable to suppose that it represented some type of bridging agent.

Biochemical work suggests that the adhesion of cells, which have been separated by trypsinisation, requires the synthesis of some component. The idea has been advanced (see Moscona, 1962) that the material synthesised is directly involved in cell adhesion. If, however, the plasmalemma has been damaged during the preparation of the cells by trypsinisation, resynthesis of a normal or quasi-normal surface will be required before adhesions can form. This re-synthesis may be the process which limits adhesion. Detailed examination of aggregation kinetics (Curtis, 1970a; Edwards and Campbell, 1971) suggests that such 'repair' processes may account for the apparent finding that cell adhesion depends upon the synthesis of protein (Moscona, 1968) or amino-sugar components (Oppenheimer et al., 1969). In many instances no metabolic requirement or involvement in adhesion has been demonstrated (Humphreys, 1963, for sponge cells; Curtis and Greaves, 1965; Glaeser et al., 1968, for avian embryonic cells; Salt, 1965 for the adhesion involved in the haemocyte reaction to foreign bodies and parasites in insects). It is of interest that in these instances the cells were not exposed to trypsin whereas, when evidence for a process of synthesis was obtained, the cells had been exposed to this enzyme. If there was definite evidence in favour of the idea that the synthetic step is directly involved in producing the adhesion, this would be a strong reason for accepting the bridging theory. Since, in fact, it looks as though the evidence for a synthetic step can be better interpreted as evidence for repair after a special method of cell preparation, it may be possible to conclude that biochemical evidence tends to support either the molecular contact or the secondary minimum theories. However, it should be remembered that one method by which one organism may prevent the attachment of another is by the release of enzymes that damage the surface of the other organism. A number of papers (e.g. Lilien, 1968; Oppenheimer and Humphreys, 1971) have reported the isolation of macro-molecular factors aiding adhesion. Though it has been usual for the authors to claim that there are the cements (bridging agents) that bind cells, no direct evidence of this has been produced (see below).

The nature of the ionic conditions that affect cell adhesion tends to support the idea that electrostatic forces of repulsion play a role in the formation of an adhesion. This is reviewed by Curtis (1967 and in press). It is clear, however, that there are a number of papers which do not report the relationships to be expected on this hypothesis. The reason for this discrepancy is unclear though it may be due to the fact that changes in conditions which affect electrostatic forces towards an increase in adhesiveness may also reduce the forces of attraction, thus cancelling or even reversing the effects to be expected.

The disaggregating effect of proteolytic enzymes such as trypsin, which is a most effective agent for separating tissues into cells, can be most readily explained on the bridging hypothesis, though it is possible to devise complex and unproven explanations on the other two hypotheses.

In conclusion it can be seen that none of the theories can yet be said to receive unequivocal support from the experimental evidence. Though it might be suggested that the discrepancies are due to the fact that very different cell types have been used in some of the experiments this is probably an incorrect explanation, because some of the most conflicting results come from studies on identical, or very similar, cell types. The DLVO theory is successful at a quantitative level in predicting the existence of two types of adhesion, the weak adhesions with a gap between surfaces and the strong zonulae occludentes. It also suggests that ionic conditions should affect adhesion and that the process should not require the synthesis of any material to provide chemical bonding between cells: these two predictions are at least partly borne out by experimental work. The bridging theory does not explain these matters, though it accounts for the effect of enzymes in separating cells more successfully than other theories. Work on the adhesion of cells to non-living substrates (reviewed by Curtis, 1967) suggests that the mechanisms involved are very similar to, if not identical with, those in cell to cell adhesion.

SYSTEMS FOR THE CONTROL OF CELL ADHESION

Despite this lack of certainty as to the nature of the mechanism of cell adhesion, experimental results and theory clearly indicate a number of ways in which adhesive relations between one cell, or one organism and another, might be controlled.

The following systems could be used for the control of adhesion:

(i) Alteration in the ionic nature of the local environment.

(ii) Alteration in the dielectric constant of the immediate surroundings of the cell.

(iii) Changes in the nature and packing of cell surface lipids.

(iv) Changes in the surface charge of the cell.

(v) The release or inhibition of enzymes that attack the adhesive mechanism directly, or affect it indirectly by altering cell surface structure.

(vi) The production or destruction of specific, or non-specific, bridging agents or the synthesis of aberrant forms of these agents that would competitively inhibit adhesion.

(vii) Synthesis of specific groupings on the surface that indirectly affect adhesion.

The first mechanism is unlikely to be used by cells or parasites as a general method of controlling adhesion because cells are not able to secrete, absorb or chelate ions to an extent which would alter the composition of the external medium appreciably, save in two cases. The first arises in a compact tissue in which the proportion of extracellular space may be small compared with the volume of the cells. Such a situation is unlikely to affect the attachment of a parasite to a tissue but might conceivably be involved in the release of the parasite. The second situation occurs if the cells secrete protons. Cell adhesion is very sensitive to small pH changes, for example the adhesiveness of chick neural retina cells rises by a factor of 1×10^4 on lowering the pH from 8·0 to 7·0 (Curtis, 1963, 1969). Many parasites produce considerable amounts of acid and this may ensure that they remain in firm adhesion with their host.

Changes in the dielectric constants of the surrounding media would affect both electrostatic forces of repulsion and London forces of attraction. An increase in dielectric constant would increase the London and the electrostatic forces, though to different extents so that one effect would not be offset by the other. However, it seems unlikely that appreciable decrease or increase could be produced by the secretory products of cells.

Alterations in cell surface lipids could be induced by any of the following mechanisms:

(i) Lysis of selected components of the cell surface. For example, exo- or endogenous phospholipases might convert plasmalemmal phosphatidyl compounds to the lysophosphatidyl compounds by releasing a fatty acid chain (see Fischer *et al.*, 1967).

(ii)   Synthetic modifications of existing plasmalemmal components, e.g. the esterification of cholesterol.

(iii)   Resynthesis of lysed components into the surface. Fischer *et al.*, have shown that normal red blood cells carry acyl transferases that re-acylate lysolecithin in the surface to phosphatidyl cholines. If the fatty acid source is changed then the nature of the cell surface lecithins may be progressively changed. For example, processes of lysis may degrade say di-oleyl lecithin to oleyl lysolecithin which might re-acylate to oleyl-stearoyl lecithin if much stearic acid were available to the system. In this way progressive changes in the degree of saturation, and the length of the hydrocarbon chains in the plasma-lemma might be produced.

(iv)   The incorporation of hydrophobic materials into the cell surface and the adsorption of surface active materials to the outside of the cell. For example, vitamin A (Dingle and Lucy, 1962) probably enters the lipid portion of the membrane and there deranges the structure to a considerable extent.

At present there is little evidence that any of these mechanisms can actually operate to affect adhesion but if the DLVO theory is correct it would be expected that all of these systems could be of effect because they would all be expected to alter the dielectric constants (over all frequencies) of the cell surface. Thus, if such effects are found it would tend to provide evidence that the DLVO theory accounts for cell adhesion. However, it is possible to explain such effects on the bridging theory, for alteration in lipid structure might change the conformation and exposure of surface binding sites.

Fischer *et al.* (1967) found that red blood cells produce lysolecithin on incubation in a simple saline medium. They observed that the adhesiveness of the cells appeared to decrease after this incubation. Curtis (in preparation) found that the adhesiveness of embryonic chick cells was greatly diminished by treatment with phospholipase A, lysolecithin, or by the incubation of the cells at 37°C in simple media, which allow the release of fatty acid from the cell surface. Stimulation of the re-acylation system restored the adhesiveness of the cells unless the fatty acid incorporated was of short chain length or showed a considerable degree of unsaturation. It appears that turn-over of these cell surface acyl groups is very rapid. Treatment with antibodies and drugs of the dipyramidole series appear to block the acyl transferase system while leaving the lysolphospholipase reaction unaffected or even at an enhanced level. This is of particular interest because Dintenfass (1970) reports that a dipyramidole derivative lowers the adhesiveness of red blood cells. Sulphydryl-blocking reagents were reported by Fischer *et al.* (1967) as inhibiting both enzymes. Grinnell and Srere (1971) found that such reagents block the adhesiveness of cells, which might be expected if the re-acylation was a necessary precursor of cell adhesion. These experiments, which are still at an early stage, suggest that control of the lipid state of the cell surface could be of importance in the control of cell adhesion.

A variety of mechanisms might act to produce changes in the surface charge of cells such as the stimulation of addition of neuraminic acid components to glycoprotein or glycolipid chains. Similarly, if one cell or organism could release a specific glycosidase it might be able to alter the surface charge of another cell thus specifically controlling a non-specific mechanism of adhesion. Unfortunately, the comparatively small amount of work carried out to measure the surface charge of cells in relation to their adhesiveness has been largely vitiated by two matters. The first of these has been the failure of most workers to use any of the semi-quantitative methods of measuring adhesion to test for a correlation. Fully quantitative methods of measuring adhesion have only become available very recently (Curtis, 1969). The second difficulty lies in the fact that it has been usual to carry out electrophoresis of cells (to measure their surface charge) in simple media of the type which cause rapid loss of lecithins due to inhibition of the system for re-acylating lysolecithin. These changes in the cells may tend to affect surface charge thus possibly abolishing differences that exist between cells in a more normal environment.

Several of the ways in which enzymic action might modify adhesiveness have already been considered. These examples are mostly ones in which the enzymes lyse a component involved either directly or indirectly in cell adhesion. Adhesion might, however, be controlled by the enzymic synthesis of cell surface components. If adhesion is achieved by a bridging system it is only necessary to suppose that the synthesis attaches the agent to one or both surfaces. Roseman (1970) has proposed an ingenious scheme for the role of surface glycoproteins in cell adhesion. He suggests that the glycosyl transferases on one cell surface catalyse the addition of amino-sugars to the growing glycosidic chains of the glycoprotein on an adjoining cell surface. Thus at one stage of each addition the two surfaces are bound together by the enzyme substrate reaction. If the enzyme is not released from the amino-sugar after its addition to the chain is complete, the adhesion would become fairly permanent. Such a mechanism would provide a labile but possibly highly specific means of sticking one cell to another. There are, however, a number of difficulties with this hypothesis. Calcium ions are often inhibitors of glycosyl transferases as Roseman points out. In consequence, calcium might be expected to be an inhibitor of adhesion but on the whole it appears to be a factor promoting adhesiveness according to experimental evidence. Second there is the possibility that cells of unlike type might bear amino-sugar sequences on their surfaces which irreversibly inhibit the transferases by binding to them. Thus it might be expected that unlike (heterotypic) cells would adhere more strongly than like cells. The third difficulty with the theory is that it would be necessary to have roughly the same density and spacing of enzyme molecules and glycoproteins on a cell surface to ensure that adhesion was at all efficient and of any appreciable strength. It seems unlikely that the enzymes are as numerous as the glycoprotein molecules. Clegg *et al.* (1971) have found that host antigens may be transferred to the parasite surface. It is interesting to speculate whether host glycosyl transferases might be able to synthesise host antigens on the parasite. This is an alternative to the concept of the transfer of whole glycolipid molecules from host to parasite cell surface suggested by Clegg and colleagues.

It can be seen that, so far, little work has been carried out in this field but that there are a fairly large number of ways in which cells might control not only their own adhesiveness but that of other neighbouring cells as well. Some of the potential systems may not, in fact, be used by organisms but may provide ideas as to methods of designing systems to control adhesion in various form of disease and parasitic infection. I mentioned earlier that the evidence for the involvement of a synthetic step in the formation of an adhesion could be better interpreted as 'repair' of the surface, rather than the formation of a bridging agent. For this reason the effect of various reagents in diminishing adhesion may lie not in the fact that they destroy a specific bonding system but rather that they damage the cell surface so that it cannot use another system of adhesion. For example, trypsinisation of cells may damage the acyl transferases so that maintenance of the normal level of phosphatidyl cholines is impossible. As a result the cell loses adhesion, perhaps because the dielectric constant of the surface rises. Thus it could be erroneous to make deductions about the mechanism of adhesion from observations of those factors that affect the adhesiveness of cells. However, it should also be borne in mind that the actual control systems used by cells and organisms could act both directly or indirectly, as trypsin may, on the adhesive mechanisms of themselves or other cells.

SPECIFIC SYSTEMS FOR ADHESIVE INTERACTION

It has often been suggested that the mechanism of cell adhesion shows a considerable degree of specificity (e.g. Wilson, 1907; Moscona, 1962). Clearly this statement cannot be wholly correct because apparently all types of adhesive cell will adhere to a very wide range of non-living substrates, such as various glasses, silica, a wide range of plastics including polyethylenes, polytetrafluorethylenes, acrylates, nylons etc., metals such as stainless steels, tantalum, gold and platinum, and many fibrous proteins etc. Moreover it seems probable that these adhesions have basically the same properties as those formed between cells (Curtis, 1967). However, experiments such as those of Wilson (1907) have been used to argue that cells normally show specific adhesion. Wilson found that when cell suspensions from two different sponge species were mixed, each single aggregate that resulted appeared to be composed of the cells of one of the species alone. This result could be explained by many different mechanisms (Curtis, 1967) but Wilson suggested that the cells of each species could only adhere to their own type—a process of specific adhesion. If such a mechanism is at all widespread, parasites must be able to mimic the specific adhesive processes of the host, while the elimination or release of a parasite might depend on the re-establishment of the specificity in such a way that the parasite can no longer adhere. Such possibilities are of considerable interest. It is clear that non-specific cell adhesion between cells must occur because cells of different types stick to each other in many different circumstances. Armstrong (1970) has shown that such contacts have the same appearance by electron microscopy as those between cells of the same type.

Wilson's observations are not sufficient to prove the occurrence of specific adhesion because a number of other explanations cannot be excluded. For example Steinberg (1963) has described a mechanism whereby quantitative differences in the adhesiveness of two cell types might account for their separate adhesion. Curtis (1967) proposed a system of timed differences in adhesiveness that would produce the same effect. The apparent homing of primordial germ cells and lymphocytes to specific sites in the vertebrate body have been claimed as examples of specific adhesion. However Meyer (1964) has shown that primordial germ cells move and locate randomly in the body and there is reason to think that the processes whereby lymphocytes accumulate in certain regions of the body are remarkably unselective (Curtis, 1967). The sorting out (segregation) of cells of different types from the same species, which takes place within an aggregate body soon after reaggregation, has been explained by Moscona (1962) as being due to the operation of specific adhesion. Both Steinberg (1964) and Curtis (1962) have pointed out that such a mechanism would produce random arrangement of groups of cells of one type, whereas the position of cell types after sorting out is highly patterned. Thus it appears that there is little support for the concept of specific adhesion from these phenomena. However, these experiments are of great interest because adhesive and other contact reactions between cells of different organisms can be studied by such means.

The first direct experimental test for the occurrence of specific adhesion of cells was introduced by Roth and Weston (1967), see also Roth (1968). Though this test has not been applied to the problem of whether cells of different species adhere by different mechanisms, the general approach is of great relevance to such a situation. Roth and Weston prepared aggregates of either liver or neural retina cells from embryonic chicks. These aggregates were then introduced into suspensions of radioactively labelled liver or neural retina cells. They found that labelled liver cells tended to stick to liver aggregates in much greater proportions than to neural retina aggregates, while neural retina cells adhered to retinal aggregates more readily than to liver aggregates. Specificity was not complete but it appeared that there was a very marked preference for cells to adhere to aggregates of their own type. Curtis (1970c) criticised this experiment on the grounds that the selectivity could be accounted for by the sequence of changes in adhesiveness shown by these two cell types after the start of aggregation. He showed that liver cells are very adhesive at the start of aggregation while neural retinal cells are of low adhesiveness. After about three hours of reaggregation the relative adhesiveness of the two cell types suddenly reverses. It could be demonstrated that the surface of the aggregates showed a similar change in adhesiveness. Therefore the neural retina cells and aggregates would be adhesive at a time when the liver cells were of low adhesiveness and vice-versa. Such a mechanism of 'temporal specificity' would ensure that, when two different organisms were in competition for an attachment site, an apparent specificity of attachment would develop. Such a mechanism does not imply that the adhesion itself is specific.

A quantitative method for testing for the occurrence of specific adhesion was introduced by Curtis (1970a). This is based on the method of measuring collision efficiency i.e. the probability that a cell or an aggregate adheres to another on making a collision with it, which I described recently (Curtis, 1969). If two cell types of equal collision efficiency E are mixed in any proportion, the value of E_{mix} obtained will be the same as that for either type alone, provided that there is no specificity of adhesion. If, however, there is complete specificity of adhesion the value of E_{mix}, measured for a mixture of equal parts of each type of cell, will be half that for either type alone because half the collisions will be between cells and aggregates of the opposite type (which are by definition of complete specificity, ineffective). In general the value of E which will be expected for a mixture of cells in proportions n_1 and n_2 of collision efficiencies E_1 and E_2 is given by:

$$E_{mix} = n_1 E_1 + n_2 E_2$$

for non-specific adhesion and by:

$$E_{mix} = n_1^2 E_1 + n_2^2 E_2$$

for complete specificity of adhesion. Thus measurement of the collision efficiencies of pure cell types and of their mixtures provides a method for detecting whether specific adhesion occurs, and for recognising situations in which there is either a partial degree of specificity or in which only part of the cell population displays specificity, see also Fig. 3.

The method was used by Curtis (1970a, b) to investigate whether specific adhesion took place in various species of marine sponge. Cells from *Haliclona occulata, Halichondria panicea, Microciona fallax* and *Suberites ficus* were used and he found that well washed cells showed no evidence of specific adhesion, in all the pair-wise combination tested. The same method was used by Dr van de Vyver and Curtis (Curtis and van de Vyver, 1971) to investigate whether

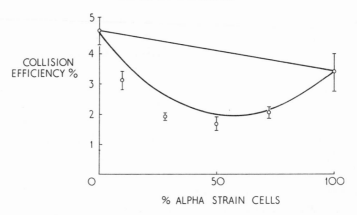

FIG. 3. Test for the specific adhesion of cells. The collision efficiency (adhesiveness) of the mixed cells is plotted against the percentage of cells of one type in a mixture. The upper line shows the results that would be expected on the hypothesis that specific adhesion does not occur, the curve shows those anticipated if completely specific adhesion takes place. The actual experimental points refer to mixtures of alpha and delta strains of the sponge *Ephydatia fluviatilis* (see Curtis and van de Vyver, 1971). In this case the experimentally established points deviated from both expectations so that evidence was obtained for another system of control of adhesion (see text).

specific adhesion took place in cells of the different strains of the freshwater sponge *Ephydatia fluviatilis*. The different strains of this species are recognised by the phenomenon of 'non-coalescence', namely that when two sponge bodies of different strain type are placed side by side they first fuse but then separate, leaving on occasion a zone of non-adherent cells between them. Sponges of the same strain type form a permanent fusion if placed side by side. Preliminary work using the collision efficiency test suggested that the cells of these strain types might show specific adhesion but further investigation showed that the mechanism of adhesion was non-specific. However the strains produce soluble and probably diffusible substances that act on cells of heterologous strains to diminish their adhesion while increasing that of cells of the homologous strain. This behaviour accounts for the apparent occurrence of specific adhesion and explains the phenomenon of non-coalescence. The most important feature of this system is the specific control of the quantitative value of the adhesion while the adhesion itself takes place by a non-specific mechanism. At present little is known about the nature of these factors.

Oka's work (Oka, 1970) on *Botryllus* colonies shows considerable parallel with the phenomenon of 'non-coalescence'. Organisms which are partially or wholly isogenic with one another show fusion. Unlike organisms will not form permanent fusions (it is unclear as to whether an impermanent fusion may be formed or not) and a zone of loose unadherent and possibly necrotic cells appears between the colonies. Burnet (1971) has suggested that this phenomenon is a primitive example of a self-recognition process by which a positive recognition occurs and which is 'almost certainly correctly interpreted as representing specific union, reversible or irreversible, between chemical groupings on the surface of the interacting cells'. There is as yet no evidence in favour of such a conclusion and it seems possible that the *Botryllus* system resembles that in *Ephydatia* in which the specificity resides in the control of the strength of the adhesion and not in the mechanism of adhesion itself.

Theodor (1970) has shown that similar contact interactions take place both in xenogenic contacts between the gorgonians *Lophogorgia sarmentosa* and *Eunicella stricta* and between allogenic contacts of individuals of either species. These interactions lead to disintegration of the tissues at and near the site of contact. It is unclear at present whether cell adhesion is affected and whether cytolysis occurs. Theodor suggested that this phenomen could be explained as an example of self-recognition processes of a quasi-immunological nature. However it can also be interpreted as an example of the type of adhesive system found in *Ephydatia*. This raises the general question of how far specific systems of the control of adhesion (rather than specific adhesion) can be classified as self-recognition systems, for they form a class whose existence does not appear to have been recognised previously and which are apparently considerably different in mechanism from immunological systems.

It is clear that if two organisms are so similar that they share the same adhesion controlling factors then it would be expected that they could fuse, or one could become epizootic or even parasitic on the other. An alternative road to parasitism is for the two organisms to differ so much in their adhesion controlling systems that the parasite is unaffected by any adhesion controlling factors released by the host. This would be achieved by ensuring that the parasite had cell surfaces or other external coverings whose chemistry differed very considerably from that of the host. The first system would lead to the development of a particular association between two species, whereas the second method would tend to favour epizooism or parasitism by a whole range of organisms unrelated to the host. It is of interest that Rutzler (1970) has found that epizoism in sponges tends to occur between certain pairs of species.

Processes leading to the specific localisation of the attachment of organisms also take place so that epizooisms and parasitism are confined to particular regions of an organism. These localisations may represent the action of particular adhesive relationships at certain parts of the surface of the host. Specific localisation even occurs at the cellular level. Dr Meadows (personal communication) has investigated the attachment of bacteria to cells in culture and has found that attachment may tend to be confined to the leading edge (lamellopodium) region of a fibroblast (see Fig. 4). At present it is unclear as to whether a simple quantitative increase in adhesiveness in this region, the presence of specific binding sites, or the operation of short-range diffusible factors effects this localisation.

W.I.38 – Human Embryonic Lung

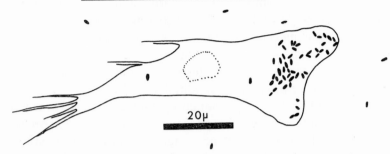

FIG. 4. The non-random distribution of bacteria adhering to a cell in tissue culture. By kind permission of Dr P. Meadows.

In conclusion it appears that the development of specific associations between cells or organisms is not due to a system of specific adhesion for, with the exception of Crandall and Brock's (1968) work on yeast mating types, no clear evidence for the occurrence of specific adhesion has yet appeared. Moreover, some of the systems whose behaviour has suggested the action of specific adhesion are better explained by other theories. The concept of specific diffusible factors (affecting the quantitative level of adhesion and this in turn determining whether an adhesion forms or not by a non-specific mechanism) explains a number of systems for which specificity of association has been claimed, more satisfactorily. A number of papers (Lilien, 1968; Humphreys, 1965; Oppenheimer and Humphreys, 1971; Moscona, 1968) have described the isolation of factors that promote adhesion. These papers suggest that the factors form specific cements (bridging agents) between cells but this has not been demonstrated. It would be of interest to test the effects of these factors on heterologous cells to discover whether they diminish their adhesiveness in the manner of the *Ephydatia* factor. The *Ephydatia* system provides a means of preventing epizoism or parasitism by a closely related organism, a question for the future is whether such systems control the development of parasitism in other organisms.

REFERENCES

ARMSTRONG, P. B. (1970). A fine structural study of adhesive cell junctions in heterotypic cell aggregates. *Journal of Cell Biology* **47**: 197–210

BRETSCHER, M. S. (1971). Major human erythrocyte glycoprotein spans the membrane. *Nature, London* **231**: 46–52

BRIGHTMAN, M. W. (1965). The distribution within the brain of ferritin injected into cerebrospinal fluid compartments. I. Ependymal distribution. *Journal of Cell Biology* **26**: 99–123

BURNET, F. M. (1971). 'Self-recognition' in colonial marine forms and flowering plants in relation to the evolution of immunity. *Nature, London* **232**: 230–35

CLEGG, J. A., SMITHERS, S. R. and TERRY, J. R. (1971). Acquisition of human antigens by *Schistosoma mansoni* during cultivation *in vitro*. *Nature, London* **232**: 653–4

COOK, G. M. W., SEAMAN, G. V. F. and HEARD, D. H. (1960). A sialomucopeptide liberated by trypsin from the human erythrocyte. *Nature, London* **188**: 1011–2

CRANDALL, M. A. and BROCK, T. D. (1968). Molecular aspects of specific cell contact. *Science* **161**: 473–5

CURTIS, A. S. G. (1962). Cell contact and adhesion. *Biological Reviews* **37**: 82–129

CURTIS, A. S. G. (1963). The effect of pH and temperature on cell reaggregation. *Nature, London* **200**: 1235–6

CURTIS, A. S. G. (1967). *The Cell Surface: its Molecular Role in Morphogenesis*. Pp. x + 405. London: Logos Press, Academic Press

CURTIS, A. S. G. (1969). The measurement of cell adhesiveness by an absolute method. *Journal of Embryology and experimental Morphology* **22**: 305–25

CURTIS, A. S. G. (1970a). Problems and some solutions in the study of cellular aggregation, *Symposia of the Zoological Society of London* **25**: 335–52

CURTIS, A. S. G. (1970b). Re-examination of a supposed case of specific cell adhesion. *Nature, London* **226**: 260–1

CURTIS, A. S. G. (1970c). On the occurrence of specific adhesion between cells. *Journal of Embryology and experimental Morphology* **23**: 253–72

CURTIS, A. S. G. (1972a). Intra- and inter-membrane interactions of the cell surface. *Subcellular Biochemistry* (in press)

CURTIS, A. S. G. (1972b). General functions of the cell surface. From *Cell Biology in the Service of Medicine*. E. Bittar (ed.). Wiley & Co.

CURTIS, A. S. G. and GREAVES, M. F. (1965). The inhibition of cell aggregation by a pure serum protein. *Journal of Embryology and experimental Morphology* **13**: 309–26

CURTIS, A. S. G. and HOCKING, L. M. (1970). Collision efficiency of equal spherical particles in a shear flow. *Transactions of the Faraday Society* **66**: 1381–90

CURTIS, A. S. G. and VAN DE VYVER, G. (1971). The control of cell adhesion in a morphogenetic system. *Journal of Embryology and experimental Morphology* **26**: 295–312

DINGLE, J. T. and LUCY, J. A. (1962). Studies on the mode of action of excess of Vitamin A. *Biochemical Journal* **84**: 611–21

DINTENFASS, L. (1970). Influence of ABO blood groups on the selective disaggregation of the red cells caused by drug RA 433. *Medical Journal of Australia* (2): 827–30

EDWARDS, J. G. and CAMPBELL, J. A. (1971). The aggregation of trypsinized *BHK21* cells. *Journal of Cell Science* **8**: 53–72

FARQUHAR, M. G. and PALADE, G. (1963). Junctional complexes in various epithelia. *Journal of Cell Biology* **17**: 375–412

FISCHER, H., FERBER, E., HAUPT, I., KOHLSCHUTTER, A., MODOLELL, M., MUNDER, P. G, and SONAK, R. (1967). Lysophosphatides and cell membranes. *Protides of the Biological Fluids* **15**: 175–84

FUCHS, N. (1934). Uber die stabilitat und aufladung der aerosole. *Zeitschrift für Physik* **89**: 736–43

GLAESER, R. M., RICHMOND, J. E. and TODD, P. W. (1968). Histotypic self-organization by trypsin-dissociated and EDTA-dissociated chick embryo cells. *Experimental Cell Research* **52**: 71–85

GOEL, S. C. and JURAND, A. (1971). The structures at the epithelio-connective tissue junction: a comparison of light and electron microscopic observations. *Proceedings of the Royal Society of Edinburgh* B **71**: 1–13

GRINNELL, F. and SRERE, P. A. (1971). Inhibition of cellular adhesiveness by sulfhydryl blocking agents. *Journal of Cellular Physiology* **78**: 153–7

HENDLER, R. W. (1971). Biological membrane ultrastructure. *Physiological Reviews* **51**: 66–97

HUMPHREYS, T. (1963). Chemical dissolution and *in vitro* reconstruction of sponge cell adhesion. I. Isolation and functional demonstration of the components involved. *Developmental Biology* **8**: 27–47

HUMPHREYS, T. (1965). Cell surface components participating in aggregation: evidence for a new cell particulate. *Experimental Cell Research* **40**: 539–543

KELLY, D. E. and LUFT, J. H. (1966). Fine structure, development and classification of desmosomes and related attachment mechanisms. Pp. 401–2. From *Electron Microscopy. VIth International Congress of Electron Microscopy*. Kyoto, Tokyo: Maruzen

LILIEN, J. E. (1968). Specific enhancement of cell aggregation *in vitro*. *Developmental Biology* **17**: 657–78

MEYER, D. B. (1964). The migration of primordial germ cells in the chick embryo. *Developmental Biology* **10**: 154–90

MOSCONA, A. A. (1962). Analysis of cell recombinations in experimental synthesis of tissues *in vitro*. *Journal of Cellular and Comparative Physiology* (Suppl. 1) **60**: 65–80

MOSCONA, A. A. (1968). Cell aggregation properties of specific cell ligands and their role in the formation of multicellular systems. *Developmental Biology* **18**: 250–77

NAPPER, D. H. (1967). Modern theories of colloid stability. *Science Progress* (*Oxford*) **55**: 91–109

NINHAM, B. W. and PARSEGIAN, V. A. (1970). Van der Waals forces. Special characteristics in lipid-water systems and a general method of calculation based on the Lifshitz theory. *Biophysical Journal* **10**: 646–63

OKA, H. (1970). Colony specificity in compound ascidians. Pp. 195–206. From *Profiles of Japanese Science and Scientists*. H. Yukawa (ed.). Tokyo: Kodansha

OPPENHEIMER, S. B., EDIDIN, M., ORR, C. W. and ROSEMAN, S. (1969). An l-glutamine requirement for intercellular adhesion. *Proceedings of the National Academy of Sciences of Washington* **63**: 1395–1402

OPPENHEIMER, S. B. and HUMPHREYS, T. (1971). Isolation of specific macromolecules required for adhesion of mouse tumour cells. *Nature, London* **232**: 125–7

PARDOE, G. T. and UHLENBRUCK, G. (1971). Genetics and immunochemistry of blood group antigens. *Medical Laboratory Technology* **28**: 1–18

ROSEMAN, S. (1970). The synthesis of complex carbohydrates by multi-glycosyltransferase systems and their potential function in intercellular adhesion. *Chemistry and Physics of Lipids* **5**: 270–97

ROTH, S. (1968). Studies on intercellular adhesive selectivity. *Developmental Biology* **18**: 602–31

ROTH, S. and WESTON, J. A. (1967). The measurement of intercellular adhesion. *Proceedings of the National Academy of Sciences of Washington* **58**: 974–80

RUTZLER, K. (1970). Spatial competition among Porifera: solution by epizoism. *Oecologia* **5**: 85–95

SALT, G. (1965). Experimental studies on insect parasitism. XIII. The haemocytic reaction of a caterpillar to eggs of its habitual parasite. *Proceedings of the Royal Society* (Series B) **162**: 303–18

STEINBERG, M. S. (1964). The problems of adhesive selectivity in cellular interactions. Pp. 321–66. From *Cellular Membranes in Development*. M. Locke (ed.). New York: Academic Press

THEODOR, J. (1970). Distinction between 'Self' and 'Not-self' in lower invertebrates. *Nature, London* **227**: 690–2

VICKERMAN, K. (1969). On the surface coat and flagellar adhesion in trypanosomes. *Journal of Cell Science* **5**: 163–93

VISSER, J. (1968). Adhesion of colloidal particles. *Report of the Progress of Applied Chemistry* **53**: 714–29

WEISS, L. (1967). *The Cell Periphery, Metastasis, and Other Contact Phenomena*. Pp. 380. Amsterdam: North Holland

WILSON, E.V. (1907). On some phenomena of coalescence and regeneration in sponges. *Journal of Experimental Zoology* **5**: 245–58

ZAHLER, P. H. (1969). The structure of the erythrocyte membrane. *Experientia* **25**: 449–56

THE SCHISTOSOME SURFACE IN RELATION
TO PARASITISM

J. A. CLEGG

Division of Parasitology
National Institute for Medical Research,
Mill Hill, NW7 1AA

One of the basic problems which all parasites have to overcome is the need to infect new hosts; however successfully the parasite is adapted to the internal environment of its host it must emerge at some time into the outside world and face the hazards of transferring to another animal. *Schistosoma mansoni*, like other trematodes, has two quite different species of host, a fresh water snail and a vertebrate. The larval stage which emerges from the snail, the cercaria, spends a short time in the water and then penetrates through the skin of man or, in the laboratory, a number of other vertebrates. Thus, within a very short period, the cercaria which was adapted for life within the snail must withstand a period in water and then quickly re-adapt to life inside a vertebrate; truly a phenomenal performance. The outer surface of the cercaria is particularly concerned in this rapid re-adaptation; consequently the first problem I shall discuss concerns the changes in the structure and function of the surface membrane of the cercaria as it penetrates from water into the skin of the vertebrate host.

When the parasite has gained entry into the host and adapted successfully to the new environment it has then to deal with a problem which no free-living animal encounters. The living organism which the parasite has entered eventually mounts an immune response; the environment becomes actively and specifically hostile to the invader.

Surprisingly, many parasitic worms seem able to evade the consequences of this immune response in some way and live on in the host for considerable periods. *Schistosoma mansoni* for instance can live for several years in man even though it is in direct contact with the blood. Our group at Mill Hill has been particularly interested in this phenomenon because we have obtained evidence that the parasite protects itself against the immune response by incorporating molecules with host-like antigenic determinants into its surface membrane. This is the second aspect of surface function with which I shall be dealing.

This paper is not intended as a review of what is currently known about the functions of trematode surface membranes; it is a highly selective concentration

23

on the way a schistosome deals with two particular problems of the parasitic
way of life.

THE TRANSITION FROM THE EXTERNAL TO THE
INTERNAL ENVIRONMENT

Schistosomes enter the vertebrate host by direct penetration through the skin.
The infective stage, the cercaria, escapes from the tissues of the intermediate
host snail into the surrounding water. The cercariae are not able to exist for long
in this free-living stage, within a few hours many have lost their infectivity and
they all die within about a day if a vertebrate host is not contacted (Olivier,
1966).

The surface of the cercaria has been shown by electron microscopy to be a
thin acellular layer of cytoplasm, varying in thickness from 0·1–0·5 μm and
bounded at its outer surface by a typical trilaminate plasma membrane (Plate 1;
Smith et al., 1969; Morris, 1971). There are no nuclei in this surface layer but
thin strands of cytoplasm run down to cell bodies each containing a nucleus
lying in the parenchyma below the muscle layers (Fig. 1). This structure is a
simple version of the so-called tegument which electron microscopists have
demonstrated in a number of adult trematodes and cestodes (Lee, 1966). Until
the advent of electron microscopy, trematodes and cestodes were believed to
have a thick inert protective surface cuticle but in fact the surface is a delicate
layer of cytoplasm apparently quite unsuitable for protection of the organism.

In the cercaria the plasma membrane on the outer side of the tegument is
covered by an additional surface coat (Fig. 1; Plate 1). After preparation for
electron microscopy this surface material, which is 0·25–0·5 μm thick, appears
to be fibrillar. The fibres are about 15 Å thick and although orientated primarily
at right angles to the surface membrane they are branched and interconnected
to form a diffuse network (Morris, 1971). Histochemically the surface coat is
known to give a strong periodic-acid-schiff (PAS) reaction indicating the
presence of 1,2-glycol groups (Stirewalt, 1963; Smith et al., 1969), and the use
of colloidal iron has shown that the material also contains acidic groups (Smith
et al., 1969). The coat is thus considered to be composed primarily of acid
mucopolysaccharide, but it is worth noting that these reactions can also be given
by lipoproteins containing carbohydrate (Clegg, 1965).

The surface coat of the cercaria is of particular interest because most of it is
lost when the cercaria penetrates into the skin of the vertebrate host and
becomes a schistosomulum. Electron microscopy of the schistosomula, recovered from the lungs of mice a week after infection shows clearly that the
surface coat has largely disappeared (Smith et al., 1969). Hockley (1970) has
shown that the surface coat is actually lost within the first 15 minutes after

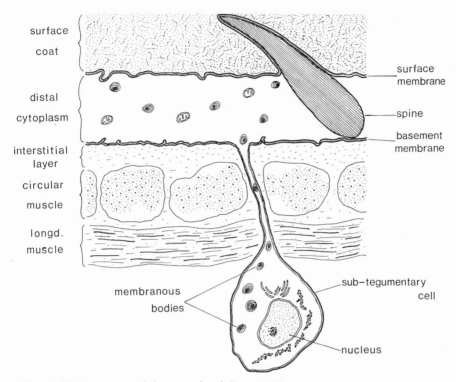

surface
coat

distal

cytoplasm

interstitial
layer

circular

muscle

longd.
muscle

surface
membrane

spine

basement
membrane

sub–tegumentary
cell

membranous

bodies

nucleus

FIG. 1. The tegument of the cercaria of *S. mansoni*.

The membranous bodies are thought to be synthesised in the subtegumental cells and move up into the distal cytoplasm to fuse with the surface membrane. Diagram based on electron micrographs of Morris (1971) and Hockley (1970).

penetration; a conclusion also reached by Stirewalt (1963) on the basis of histochemical observations with the light microscope. However, Stirewalt considered that the surface had been modified rather than completely lost because a thin film of PAS positive material still remained at the surface of the schistosomulum after penetration. Electron microscopy confirms this view; a thin fibrillar coat is retained on the surface membrane of schistosomula (Smith *et al.*, 1969; Hockley, 1970).

The loss of most of the surface coat coincides with a major change in the physiology of the cercaria. Schistosomula recovered from mouse skin 15 minutes after penetration die rapidly if returned to water, the medium from which the cercariae had penetrated into the skin. This sudden sensitivity to water does not occur in all schistosomula at the same time; sequential recoveries of large numbers of schistosomula from mouse skin have shown that about 60% cannot be killed by water 15 minutes after penetration, about 10% have still not become sensitive after 30 minutes but after an hour all the organisms have undergone the change (Clegg and Smithers, 1968).

The correlation between the loss of most of the surface coat and the sensitivity of the schistosomula to water suggests that the coat material is concerned with controlling the permeability of the membrane (Stirewalt, 1963; Morris, 1971). A relatively impermeable surface coat on the cercaria could prevent outward loss of ions and also control the uptake of water. Removal of most of the coat from the schistosomula during penetration would allow free exchange when the organisms had entered the tissue fluids. However, a return to water after loss of the surface coat would result in rapid uptake of water and death.

Initially it must be necessary for the schistosomula to absorb salt irons from the tissue fluids. The cercaria develops in a freshwater snail and almost certainly has a similar osmotic pressure which is equivalent to a freezing point depression (Δ) of about $-0.25°C$ (Robertson, 1964). The vertebrate tissue fluids surrounding the schistosomula have an osmotic pressure at least twice as great ($\Delta = -0.57°C$); consequently a sudden increase in the permeability of the surface membrane would result in a serious loss of water if salt ions were not rapidly absorbed.

Another line of evidence indicates that the surface membrane is undergoing considerable changes during the initial period of penetration. The surface of schistosomula recovered soon after penetration through isolated human epidermis is much less easily dissolved by a number of reagents than the cercarial surface (Kusel, 1970). For instance, 8M urea rapidly disintegrates the entire surface of the cercariae (the surface coat and the plasma membrane) but has very little visible effect on the surface membrane of schistosomula.

This change in the stability of the membrane must be due to a change in the nature of the chemical bonds holding together the complex three-dimensional matrix of the membrane. Concentrated solutions of urea probably dissolve proteins by breaking hydrogen bonds, which implies that the surface membrane of the schistosomula has rearranged its proteins so that more stable ionic linkages are formed. Alternatively, the membrane may be stabilized by the incorporation of ions from the tissue fluids and it is interesting that the addition of calcium ions prevents 8M urea from dissolving the cercarial surface (Kusel, 1970).

The way in which the surface coat of the cercaria is lost during penetration is quite unknown. It may be scraped off mechanically as the organism burrows through the keratin layers of the epidermis or it may be removed by one of the battery of enzymes which the cercaria secretes from its 6 pairs of penetration glands as it effects entry. Among the penetration enzymes which are no doubt primarily concerned with lysis of the skin there is known to be a protease, a carbohydrate splitting enzyme and a lipase (Stirewalt, 1966).

There is some evidence which can be interpreted to support the idea of enzymatic rather than mechanical removal of the surface coat. In serum from an immune host a thick envelope forms around cercariae; this is the cercarienhüllenreaktion (CHR) of Vogel and Minning (1953). By labelling the γ-globulin of immune serum with ferritin and using electron microscopy, Hockley (1970) showed that this envelope is due to combination of antibody with the material of the surface coat. Although the surface coat is not entirely removed during penetration, schistosomula no longer give a CHR envelope in immune serum (Stirewalt, 1963). If the coat was simply scraped off the remaining molecules would not be chemically modified but enzymatic removal would involve breaking individual molecules and might well cause loss of the antigenic determinants needed for formation of the CHR envelope.

Whatever the mechanism of removal of the surface coat the short period during which the cercaria is penetrating the epidermis is quite crucial for the organism. Recovery of large numbers of schistosomula from the skin of normal laboratory hosts within 15 minutes after penetration has shown that a substantial proportion are damaged or dead (Clegg and Smithers, 1968). The surface membrane of the damaged schistosomula apparently becomes more permeable to dyes such as eosin Y or methylene blue which are normally excluded by living cells. The percentage of schistosomula damaged in this way varies in different normal hosts; in mice about 30% do not survive the initial barrier, in rats the losses are as high as 50% but in hamsters only 10% are damaged. The cause of these sizeable losses of schistosomula in the first few minutes after penetration of the skin is not clear.

The failure of normal permeability in the surface membrane could be caused by mechanical or enzymatic damage during removal of the surface coat. Alternatively, since the maintenance of normal permeability in membranes is energy dependent, the failure of the schistosomula membrane could be caused by exhaustion of the endogenous energy reserves of the organism. The cercaria spends a variable period in water without feeding and then uses an enormous amount of energy while penetrating into the skin. The level of glycogen drops from 42 μg in 10,000 cercariae to only 2·5 μg in 10,000 schistosomula, measured shortly after penetration through isolated skin (Bruce et al., 1969). It would not be surprising if many schistosomula completely exhausted their glycogen reserves before uptake of nutrients from the tissue fluids could begin.

THE FORMATION OF A MULTILAMINATE
SURFACE MEMBRANE

The surviving schistosomula move on into the dermis and at this point, about three hours after penetration, a remarkable change occurs in the structure of the surface membrane of the schistosomula. The plasma membrane of the cercaria, lying immediately below the surface coat, has the trilaminar structure of a typical unit membrane (Morris, 1971). Electron micrographs of schistosomula recovered from skin only three hours after penetration show a surface membrane with a multilayered structure (Hockley and McLaren, 1971). As Fig. 2 shows, the surface membrane of a schistosomula has 5 layers. The simplest explanation of this change in structure is that a second unit membrane has been closely applied to the original unit membrane of the cercaria. This interpretation is supported by the appearance of the surface in the electron microscope. The central dark layer is relatively thicker and probably represents the opposed surfaces of two unit membranes.

The origin of this additional membrane is intriguing because it appears so soon after penetration. Apparently membranes stored in special vesicles, which are clearly visible in the tegument, move up to the surface and are bonded with the original membrane (Hockley and McLaren, 1971).

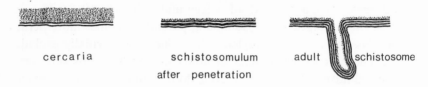

cercaria   schistosomulum   adult   schistosome

after   penetration

FIG. 2. Diagram showing the multilaminate nature of the surface membrane of schistosomula and adult schistosomes.

The free swimming cercaria has a typical trilaminar surface membrane with a thick fibrillar outer coat. Three hours after penetration into skin the schistosomulum has developed a surface membrane with 5 layers. The surface coat is nearly all removed during the very early stages of penetration. As the worm matures the surface membrane increases in complexity and becomes deeply infolded: in the adult the membrane has 7 layers (based on data from Hockley and McLaren, 1971).

There is no evidence available on the function of this change in the membrane but the fact that it takes place so soon after penetration suggests that it has an important role to play. The initial doubling of the membrane is in fact only the beginning; by the time the schistosomulum has grown into an adult worm the surface membrane has become even more complex and has seven layers (Fig. 2; Plate 6) (Smith *et al.*, 1969; Hockley and McLaren, 1971). High power electron micrographs of the adult surface often suggest that small segments of membrane are being sloughed off into the blood stream (Hockley, 1970). Together with the presence of vesicles containing rolls of membrane in the tegument (Plate 6) this evidence suggests that the surface membrane is being continuously replaced from within the tegument and lost from the upper surface. If this is so, the doubling in the thickness of the membrane which occurs in the schistosomula indicates that the process of continuous replacement begins almost as soon as the organism enters the host. Continuous replacement of the membrane could be a device to maintain a continuous supply of some labile component within it such as a transporting enzyme or it could be necessary because the membrane is being constantly damaged by an immune reaction.

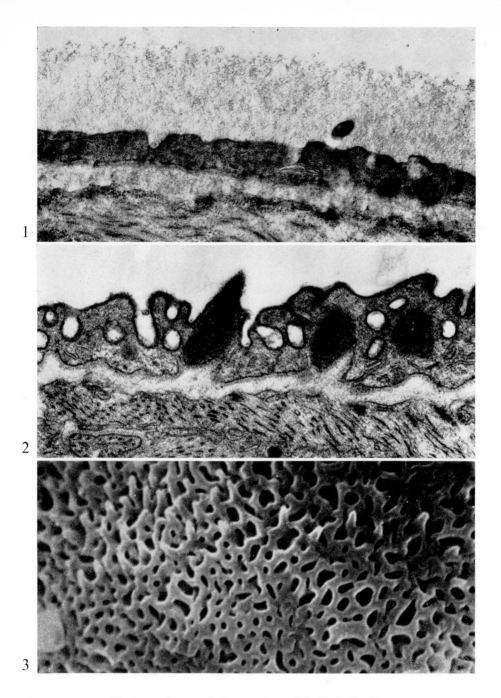

Electron micrographs by courtesy of Dr D. J. Hockley

PLATE 1. Electron micrograph of *S. mansoni* cercaria tegument. Note the fibrillar surface coat. × 22,500

PLATE 2. Electron micrograph of 7-day old schistosomulum tegument. The surface coat has been lost and the surface membrane is invaginated. × 26,500

PLATE 3. Scanning electron micrograph of 7-day schistosomulum showing the pits in the surface. × 9,000

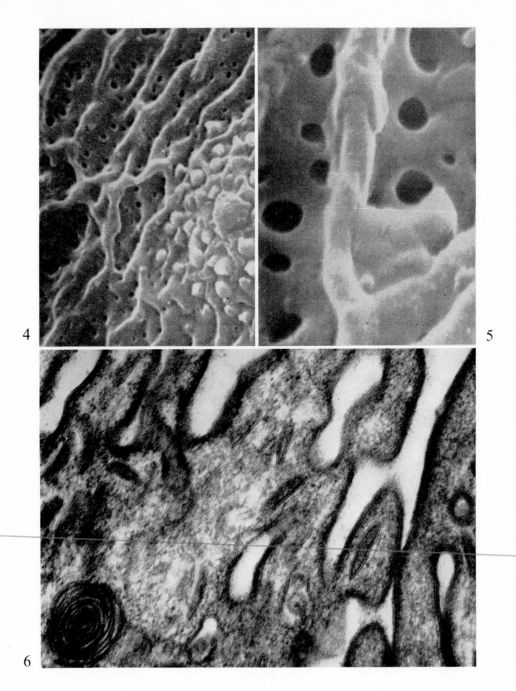

Electron micrographs by courtesy of Dr D. J. Hockley

PLATE 4. Scanning electron micrograph of adult male *S. mansoni* surface showing ridges, spines and pits. $\times$ 9,000

PLATE 5. Scanning electron micrograph—detail of surface. $\times$ 36,000

PLATE 6. Electron micrograph of adult male tegument showing the deep pits, the multilaminate surface membrane and a membranous body in the tegument which is believed to fuse with the surface membrane. $\times$ 80,000

Fixation: gluteraldehyde, osmium tetroxide and uranyl acetate.

Whatever the reason it is certainly of great interest that the surface membrane of the worm should be so complex.

The surface membrane not only becomes multilaminate but also increases very greatly in surface area during the development of the schistosomulum into an adult worm. Within the first 4 days, while the schistosomula are still within the skin or migrating to the lungs, the membrane becomes invaginated at many points. The pit-like nature of these infoldings of the membrane can be clearly seen by electron scanning of the surface of developing schistosomula (Plate 3). As the pits are forming the distal cytoplasm of the tegument is getting deeper and in the adult worm it is 4–5 μm in depth, some 10 times thicker than the tegument of the cercaria (Smith *et al.*, 1969). By this time the pits themselves are much deeper, running through the distal cytoplasm like tortuous canals (Plates 4, 5, 6). This enormous increase in the surface area of the membrane in contact with the blood must surely indicate an absorptive function. Autoradiography of sections of worms after incubation in medium containing ^{14}C-proline suggested that the tegument is an important region of absorption for this amino acid (Senft, 1968). There is also evidence of pinocytosis by the worms, since horse-radish peroxidase was able to enter the dorsal tegument of male schistosomes (Smith *et al.*, 1969). However, there have been no studies on the active transport of substances across the tegument of trematodes. This is undoubtedly due to the fact that tramatodes possess a gut; consequently radioactive nutrients entering the worm cannot be assumed to have passed through the tegument. Such a simplifying assumption can be made with cestodes and active transport has been extensively studied in these parasites. In spite of the technical problems of studying active transport in schistosomes, it is to be hoped that some way round these difficulties will be found because there is a great need to understand the function of the extraordinary multilaminate membranes of these organisms.

PROTECTION OF THE SURFACE MEMBRANE AGAINST THE IMMUNE RESPONSE

Perhaps the most striking aspect of the relationship between parasitic worms and their hosts is that they so often succeed in remaining alive long after other foreign invaders would have been destroyed by the immunological reactions of the host. One explanation could be that these large parasites manage in some way to avoid stimulating an immune response at all. In the case of *Schistosoma mansoni* this is certainly not true. In the rhesus monkey for instance, a small initial infection with *S. mansoni* stimulates a solid immunity which prevents any further reinfection by cercariae (Smithers and Terry, 1965). However, most of the adult worms derived from the original immunizing infection live on in the face of this immunity for a considerable period. This longevity of adult worms

is an important feature of the disease in man but much less is known about the
immune status of human beings harbouring schistosomes. In rhesus monkeys it
has additionally been shown by transferring adult worms from other hosts
into normal monkeys that the adult is primarily responsible for stimulating the
immune response (Smithers and Terry, 1967). Transfers of large numbers of
eggs or infection with irradiated cercariae which did not develop beyond early
schistosomula, confirmed the importance of the adult worm as the primary
stimulus to immunity. There is thus the paradox of the adult worm being unaffected by the immunity it engenders; the immune response acts mainly against
the challenging cercariae of a second or subsequent infection. This situation has
been called concomitant immunity (Smithers and Terry, 1969) following the
previous use of the term to describe an analogous phenomenon in some experimental tumours where the host animal with a particular tumour becomes
resistant to further grafts of the same cell line but does not succeed in destroying
the primary tumour.

The concept of concomitant immunity raises the basic question: how does
the adult parasite evade the effects of an immune reaction which is capable of
killing off the invading cercariae of new infections? Smithers et al. (1969) put
forward the idea that the surface of the adult schistosome becomes coated with
a molecule which has host-like antigenic determinants. This host antigen was
seen as protecting the worms against the efferent limb of the immune response
(i.e. against the effects of antibodies or sensitized cells), by masking, in some
manner, susceptible parasite antigens on the surface membrane. In this way the
immune reaction would be blocked by a molecule to which the host would be
immunologically unresponsive. Young developing schistosomula which had
not had time to adopt this disguise would be vulnerable to the immune attack.

The idea that parasite and host might share some antigenic determinants
was originally suggested by Sprent (1959). Sprent's view was that the parasite
had gradually evolved some antigens with less and less disparity from those of
the host, eventually resulting in a parasite which the host's immune system
would find difficulty in detecting. The actual existence of such common antigens
in schistosomes was then demonstrated by Damian (1964; 1967) and Capron
et al. (1965). Both these workers essentially agreed with Sprent that the parasite
had very likely developed antigens of host-type over a period of evolutionary
magnitude, which prevented the host from mounting a really effective immune
response. One shortcoming of this view of host antigen is that the host does
mount an immune response, at least against schistosomes, which prevents
re-infection. A more important objection is that a parasite genetically adapted
to a particular species of host would find the disguise quite inappropriate for
another species of host. In fact many parasites, including schistosomes, can grow
normally in a range of hosts, at least when given the opportunity to do so
in laboratory infections. To overcome this difficulty Capron et al. (1968)

suggested that schistosomes might have a series of genes coding for host antigens; the appropriate gene would then be turned on by some kind of induction process when a new host was infected.

Against this background we can see that the novel features of the Smithers and Terry hypothesis are:

(a) 'host antigens' are synthesized by the host and selectively taken up in some manner by the parasite.

(b) 'host antigens' are bound to or incorporated in the surface membrane of the schistosome.

(c) their hypothesis offers an explanation of concomitant immunity, the essential feature of which is that host antigens protect the adult worm against an immune reaction which kills invading schistosomula.

From our present standpoint the most important of these points is the second, the location of the host antigens in the surface membrane. The experimental system in which this was demonstrated involved surgical transfer of adult schistosomes from the hepatic portal system of mice into the hepatic portal system of rhesus monkeys (Smithers et al., 1969). When mouse worms were transferred into normal monkeys approximately 80% of them survived but they stopped laying eggs for about 3 weeks. Soon afterwards the worms recovered and resumed normal egg production; thus there is initial difficulty in adapting to the new host but it is only temporary. In sharp contrast, when mouse worms were transferred into monkeys previously immunized against normal mouse erythrocytes, (anti-mouse monkeys) very few survived and often all the transferred worms were destroyed. This result demonstrated that worms grown to maturity in mice have mouse antigens which render them susceptible to damage by anti-mouse monkeys. The immune reaction against 'mouse worms' is known to be mediated by antibody because normal monkeys can be passively immunized by transfer of 'anti-mouse' serum.

The mouse antigens must be firmly attached to, or even form part of, the surface membrane of the worms because the immune reaction severely damages the tegument. Electron microscopy of 'mouse worms' recovered from anti-mouse monkeys showed that the immune reaction caused the appearance of abnormal cytoplasmic bodies in the tegument within 2 hours of transfer and within 24 hours the worms were dead; in extreme cases the tegument had been entirely removed as far down as the muscle layers. The location of the mouse antigens in the surface membrane of the tegument was confirmed by electron microscopy of worms treated with ferritin-conjugated anti-mouse serum.

A criticism often raised against the demonstration of the presence of host-like antigens in the schistosome surface is that the host antigen could simply be one of a number of serum proteins contaminating the surface of the worm. Briefly, there are three lines of evidence against this view of gross contamination by serum proteins. Firstly, the antigens which schistosomes and host serum have in common are very few compared with the large number of antigenic components in serum (Damian, 1967; Capron *et al.*, 1968). Secondly, host antigen is firmly attached to the surface of the worm; washing in balanced saline does not remove it (Clegg *et al.*, 1970), and 'mouse antigens' are not removed from 'mouse worms' after three days in the circulation of a normal monkey (Smithers *et al.*, 1969). In addition, the severe damage to the surface of the worm brought about by antibody directed against host antigen clearly suggests an intimate association of the host antigen with the surface membrane. Thirdly, host antigens were not detected on all young schistosomula grown in mice for 7 days; gross contamination by serum components would surely occur very soon after the schistosomula had entered the circulation (Clegg *et al.*, 1971a).

THE NATURE OF THE HOST ANTIGENS

If host antigens are not merely gross surface contaminants from serum they may well be a serum component which is selectively taken up by the parasite. Although Smithers *et al.* (1969) immunized monkeys against mouse worms most effectively by using mouse erythrocytes, they could also immunize to some degree with mouse serum. This evidence suggests that the host antigens are present in serum and on red cells. One such component is IgG (Lahiri *et al.*, 1970) and the idea that the host antigen is actually antibody directed against the surface of the worm is attractive. Such antibodies would be firmly attached to the surface and might act to protect the worm against more damaging forms of immune response in the same way that enhancing antibodies have been shown to protect certain types of tumour cells (Uhr and Moller, 1968). Immunization of monkeys with purified myeloma mouse IgG did not support this suggestion, because 'mouse-worms' transferred into these monkeys were not affected (Clegg *et al.*, 1970). It could be reasonably argued that the enhancing antibodies in this case are present in some other class of mouse immunoglobulin. The formation of enhancing antibodies is known to be stimulated by the presence of the tumour cells in the host animal (Uhr and Moller, 1968). However, schistosomula also acquire human host antigens during 15 days culture *in vitro* in a medium containing normal human serum and erythrocytes (Clegg *et al.*, 1971b) where such stimulation of antibody formation could hardly occur. It is clear that the host antigen cannot be enhancing antibody of the kind that is known to protect some tumours.

The acquisition of human host antigens by schistosomula during cultivation *in vitro* has in fact, led to a quite different view of the nature of the molecules concerned. When schistosomula were grown for 15 days in medium containing human red cells and serum of A+ type and then transferred into monkeys immunized against washed A+ red cells, almost all the schistosomula were destroyed. A similar homologous transfer of 'B – worms' into anti-B – monkeys gave the same result. Control transfers of worms grown in 'human media' into normal monkeys, showed high levels of survival. However, when the transfer was heterologous, 'A + worms' into anti-B – monkeys, or 'B – worms' into anti-A + monkeys the percentage of worms surviving the transfer, although still much lower than in the controls, was significantly greater than in the homologous transfers (Clegg *et al.*, 1971b).

Our interpretation of these results is that the schistosomula acquired antigens *in vitro*, which were common to A + and B – erythrocytes but also, to some degree, the specific blood group antigens as well.

The A and B type antigens of human erythrocytes are known to be glyco-lipids (Fig. 3) and there are a number of other glycolipids without blood group activity in the red cell membranes and serum of all the animals so far studied (reviewed by Sweeley and Dawson, 1969).

Very recently we have carried out more direct experiments to determine whether glycolipids are host antigens (Clegg, Smithers and Terry, unpublished). We extracted the glycolipids from human red cell membranes with 95% ethanol. After treatment with chloroform and acetone to remove less polar lipids, the extract can be shown by thin layer chromatography to contain glycolipids con-taminated with 2 or 3 phospholipids and cholesterol. Monkeys were then immunized with a neutral protein, bovine serum albumin, to which the human glycolipids had been attached. When schistosomula grown in culture for 15 days in medium containing human red cells and serum were transferred into the anti-human-glycolipid monkeys, very few survived. Schistosomula grown in similar cultures survived well in control monkeys and in monkeys immunized with bovine serum albumin alone. Experiments in which adult worms grown in mice were transferred into anti-mouse-glycolipid monkeys also resulted in destruction of the worms.

These results clearly indicate that an ethanolic extract of red cell membranes contains the antigenic determinants which schistosomula acquire on their surface membrane. It is extremely unlikely that the phospholipids contaminating the glycolipid extract were involved in immunizing the monkeys because phos-pholipids have not been found to be antigenic when attached to a carrier protein (Rapport and Graf, 1969). An exception is cardiolipin, but this phospholipid does not occur in red cell membranes. On the other hand glycolipids are well known as potent lipid haptens, that is molecules too small to be antigenic but able to stimulate antibody formation when attached to a carrier protein (Sweeley and Dawson, 1969; Rapport and Graf, 1969).

At all events we can be certain that the major host antigens are polar lipids extractable from red cell membranes with 95% ethanol. Until individual glyco-lipids have been isolated from the extract we shall not be able to identify them positively as the major host antigens but they are clearly the prime candidates.

On the basis of the evidence available at present we imagine that glycolipids transfer from the serum, or from the surface of red cells via the serum, directly onto the surface membrane of the schistosome. Glycolipids have the ideal structure for this sort of manoeuver (Fig. 3). One end of the molecule which is common to all the known glycolipids present in mammalian blood consists of sphingosine with one of a number of fatty acids bonded at the amino group. This structure, known as ceramide, is very hydrophobic and glycolipids could attach themselves to lipid-containing structures, such as cell membranes, by

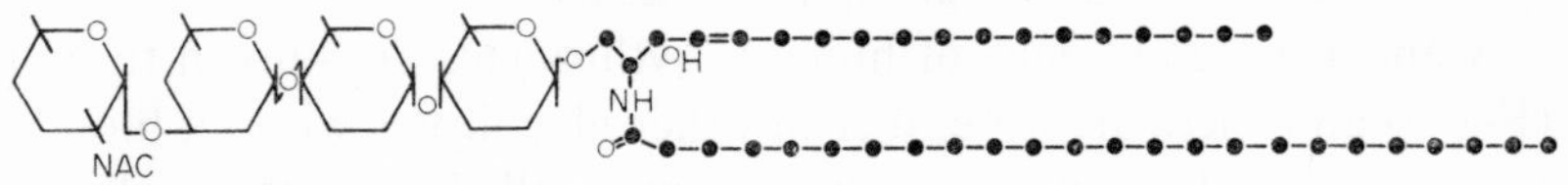

FIG. 3. A glycolipid molecule.

The hydrophobic ceramide end of the molecule, which consists of sphingosine with a fatty acid bonded at the amino group, is common to all the known glycosphingo-lipids found in mammalian blood. The other end of the molecule is a variable chain of hexose units responsible for the hydrophilic and antigenic properties of the glycolipid. The glycolipid shown here is human globoside, N-acetyl-galactosamine-(1-3)-galactosyl-(1-4)-galactosyl-(1-4)-glucosyl-(1-1)-ceramide.

interdigitating the ceramide end of the molecule between other lipid molecules. One glycolipid, the Lewis (Le^a) blood group substance, has been shown to attach to red cells which do not possess it when they are incubated in serum or saline containing the Le^a glycolipid (Marcus and Cass, 1969).

The hydrophilic property of the glycolipid molecule which gives it some degree of solubility in water is due to the chain of monosaccharides attached to the ceramide (Fig. 3). The carbohydrate moiety is usually 1–5 sugars long and in different glycolipids various combinations of glucose, galactose, fucose, n-acetyl-glucosamine, n-acetyl-galactosamine and sialic acid are known to occur. This end of the molecule is also responsible for the antigenicity of glycolipids. The glycolipid molecule has a molecular weight in the region of 1,000, much too low for it to stimulate antibody formation by itself, but when attached to a carrier protein, or a cell membrane, it acts as a hapten and can be very immunogenic.

An important advantage of this view of the way host antigens may attach to the schistosome surface membrane is that it does not imply the need for specific receptors on the membrane. The sort of membrane structure which would bind glycolipids most readily is not known but it could well be a membrane which exposed more lipid at the surface.

The available information can, of course, be interpreted in quite different ways. At present all we certainly know is that the schistosome and the red cell share antigenic determinants. These determinants on the red cells are very likely due to the carbohydrate chain of glycolipids. The same antigenic determinants on the schistosome surface membrane could be present on glycolipid synthesized by the parasite or even on a different kind of molecule such as glycoprotein. It is even possible that the antigenic determinant by itself could be transferred from the red cell membrane to a suitable receptor molecule on the schistosome membrane. Transfer enzymes are known which can convert O type human red cells into A type cells and B type cells into AB cells (Schenkel-Brunner and Tuppy, 1969). In this case a specific transfer enzyme can attach suitably activated α-n-acetyl-galactosamine to the H substance on O type cells, converting it into an A type antigen. The common factor in these alternative suggestions is that they require the schistosome to possess a remarkable ability to distinguish the nature of the host antigens.

Capron's suggestion that schistosomes have a range of genes coding for the host antigens of different host species, the suitable gene being turned on in the appropriate host (Capron *et al.*, 1968), does not explain how this recognition of the right host could be effected. Nevertheless, the question of the origin of the host-like antigens on the surface membrane of schistosomes must remain open until we have a decisive demonstration that the parasite does or does not synthesize these molecules. Identification of the chemical nature of host antigens will undoubtedly be an important step in this direction.

THE FUNCTION OF HOST ANTIGENS

Smithers *et al.* (1969), attempting to explain the basis of concomitant immunity, speculated that the host antigens on the schistosome surface might protect the adult worm against the effects of the host's immune reaction directed against the parasite. Schistosomula of a challenge infection which had not had time to acquire host antigens would be vulnerable to this immunity but the adult worms resulting from the primary immunizing infection would be protected.

This hypothesis was strengthened by the finding that schistosomula grown in mice for 15 days all possessed host antigens but not all schistosomula which had been in mice for 7 days had acquired them and cercariae did not have them at all (Clegg *et al.*, 1971a). A transfer system in which rhesus monkeys were immunized against mouse red cells was the test system employed, thus the presence of host antigen on the schistosomula was detected by antibody directed against mouse red cells. Evidently it requires more than 7 days contact with the host for all schistosomula to acquire enough host antigen to be detected in this way.

According to the hypothesis, young schistosomula without host antigen should be vulnerable to the immune reaction of the host. What then is the nature of the immune reaction? There have been many attempts to answer this question by passive transfer of immune serum or sensitized cells into normal animals. At best, transfer of immunity by these classical immunological procedures has been marginal and no light has been shed on the mechanism involved (Smithers and Terry, 1969). Recently, however, we have examined the effect of sera from hyperimmune rhesus monkeys on young schistosomula cultured *in vitro* (Clegg and Smithers, 1970; 1972). Hyperimmune rhesus serum contains an antibody of the IgG class which kills all young schistosomula cultured in its presence within three to four days. The lethal antibody is dependent on labile factors in fresh monkey serum, almost certainly complement components. Hyperimmune serum inactivated at 56°C for 30 minutes severely inhibits the growth of schistosomula *in vitro* but does not kill them. It is not yet clear whether this growth-inhibiting effect is due to lethal antibody acting without complement or a different antibody altogether. Following a small primary infection rhesus serum develops a marked growth-inhibiting property and a low level of lethal antibody at about 4 months, i.e. the time when resistance to re-infection can first be reliably demonstrated (Smithers and Terry, 1965). These antibody effects on young schistosomula may be the basis of the immune mechanism which prevents re-infection of a rhesus monkey harbouring adult worms derived from a primary infection. However, the causal rôle of lethal and growth-inhibiting antibodies in the mechanism of immunity against re-infection can only be definitely proved if passive immunization of normal monkeys by transfer of immune serum can be demonstrated.

Damage of young schistosomula by immune rhesus serum fits the concept of concomitant immunity but a more important requirement of the idea is that the schistosomula of a primary infection should become insusceptible to the immune reaction before it arises.

When schistosomula were cultured in normal monkey serum and red cells for several days before being challenged with immune serum, this is exactly what happened; after 5 days they had become completely insusceptible to the lethal and growth-inhibiting effects of immune monkey serum. A similar protective effect was observed when schistosomula recovered from the lungs of mice 4 days after infection were compared with those which had just penetrated through the skin. The 4-day old worms survived and grew normally in the presence of immune monkey serum but the early penetrants were, as usual, all destroyed (Clegg and Smithers, 1972).

It is very tempting to speculate that this protection of schistosomula against antibody is brought about by uptake of host antigens, especially since we know that human host antigens can be acquired by schistosomula during cultivation *in vitro* (Clegg *et al.*, 1971a) and by schistosomula growing in mice (Clegg *et al.*, 1971a). If we can confirm that host antigens are in fact glycolipids, the ideal experiment would be to demonstrate rapid protection of schistosomula against lethal antibody by culturing them in medium containing high levels of glycolipid. Such a demonstration would be particularly important because it would be convincing evidence of the function of host antigens in protecting schistosomula against damage by antibodies. So far the function of host antigens has been an interesting area for speculation: they certainly exist on the surface membrane of schistosomes but we have only circumstantial evidence of their function.

Assuming for the moment that these host antigens are glycolipids and that they do really protect the schistosome surface membrane against damage by antibody we can briefly look at the mechanics of how such protection might occur.

The host antigen has been located on the surface membrane of the schistosome and this must be the location of the parasite antigens which are susceptible to the immune reaction. There is some direct evidence that this is so; electron micrographs of adult worms which had been transferred into a hyperimmunized rhesus monkey (which would be capable of killing even adult worms protected by host antigen) showed damage to the surface membrane as the primary lesion (Hockley and Smithers, 1970).

If the susceptible antigen in the surface membrane, which we will call 'parasite antigen' for simplicity, is a protein, we are trying to envisage how a protein could be protected against antibody attack by molecules of glycolipid. If the small glycolipid molecule (molecular weight about 1,000) sticks to the surface membrane of the schistosome by the hydrophobic ceramide end of the molecule, it could only hinder the attachment of antibody to the parasite antigen if the major part of the parasite antigen was embedded in the structure

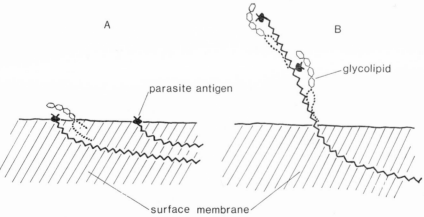

FIG. 4. Alternative attachment sites for glycolipid 'host antigens'.

(a) If the glycolipid is attached to the surface membrane by its ceramide end it could only protect an antigenic site on the 'parasite antigen' which was close to the surface.
(b) If the 'parasite antigen' sticks out some distance from the surface membrane the glycolipid would have to bind directly to the protein in order to protect its antigenic sites.

of the membrane (Fig. 4a). If the parasite antigen sticks out some distance from the surface membrane then the glycolipid molecules must attach directly to the protein in order to cover up the antigenic sites along it (Fig. 4b). This would presumably involve hydrophobic interactions between the ceramide of the glycolipid and hydrophobic regions of the protein. There is evidence that glycolipids can become attached to proteins in the surface membrane of cultured human epithelial cells (Kodani, 1962). When the cells are treated with trypsin before culture in a medium containing human A or B blood group substances, these glycolipids fail to bind to the cells. Within 48 hours, however, the cells regain their ability to absorb the glycolipids; evidently re-synthesis of the protein receptors has occurred. Whether the glycolipid protects the parasite antigen by attaching to the membrane or to the parasite antigen itself, it is not necessary to assume that all the antigenic sites on the membrane which are susceptible to damage by antibody would have to be protected. The lethal antibody is dependent on complement factors and its action on schistosomula could be prevented if it was impossible for two molecules of antibody to become attached close enough to each other to enable complement factors to be absorbed onto the attached antibody molecules.

Although there are still a number of important missing links in our view of the way the schistosome surface may be protected against the immune reaction by host antigens, we feel some confidence in the way different approaches to the problem are at last beginning to fit together.

REFERENCES

BRUCE, J. I., WEISS, E., STIREWALT, M. A. and LINCICOME, D. R. (1969). *Schistosoma mansoni*: glycogen content and utilization of glucose, pyruvate, glutamate and citric acid cycle intermediates by cercariae and schistosomules. *Experimental Parasitology* **26**: 29

CAPRON, A., BIGUET, J., ROSE, F. and VERNES, A. (1965). Les antigenès de *Schistosoma mansoni II*. Etude immunoelectrophoretique comparée de divers stade larvaires et des adultes de deux sexes. Aspects immunologiques des relations hôte-parasite de la cercaire et de l'adulte de *S. mansoni*. *Annals Institute Pasteur, Paris* **109**: 798

CAPRON, A., BIGUET, J., VERNES, A. and AFCHAIN, D. (1968). Structure antigenique des helminthes. Aspecte immunologique des relations hôte-parasite. *Pathologie et biologie, Paris* **16**: 121

CLEGG, J. A. (1965). Secretion of lipoprotein by Mehlis' gland in *Fasciola hepatica*. *Annals of the New York Academy of Sciences* **118**: 969

CLEGG, J. A. and SMITHERS, S. R. (1968). Death of schistosome cercariae during penetration of the skin. II. Penetration of mammalian skin by *Schistosoma mansoni*. *Parasitology* **58**: 111

CLEGG, J. A. and SMITHERS, S. R. (1970). The use of *in vitro* cultivation in the study of immunity against schistosomes. Proceedings 2nd International Congress of Parasitology. *The Journal of Parasitology* **56** (4) Sect. 11: 57

CLEGG, J. A. and SMITHERS, S. R. (1972). The effects of immune rhesus monkey serum on schistosomula of *Schistosoma mansoni* during cultivation *in vitro*. *International Journal for Parasitology* (in press)

CLEGG, J. A., SMITHERS, S.R. and TERRY, R. J. (1970). 'Host' antigens associated with schistosomes: observations on their attachment and their nature. *Parasitology* **61**: 87

CLEGG, J. A., SMITHERS, S. R. and TERRY, R. J. (1971a). Concomitant immunity and host antigens associated with schistosomiasis. *International Journal for Parasitology* **1**: 43

CLEGG, J. A., SMITHERS, S. R. and TERRY, R. J. (1971b). Acquisition of human antigens by *Schistosoma mansoni* during cultivation *in vitro*. *Nature (London)* **232**: 653

DAMIAN, R. T. (1964). Molecular mimicry: antigen sharing by parasite and host and its consequences. *American Naturalist* **98**: 129

DAMIAN, R. T. (1967). Common antigens between adult *Schistosoma mansoni* and the laboratory mouse. *Journal of Parasitology* **53**: 60

HOCKLEY, D. J. (1970). An ultrastructural study of the cuticle of *Schistosoma mansoni* Sambon, 1907. Ph.D. thesis, University of London.

HOCKLEY, D.J. and McLAREN, D.J. (1971). The outer membrane of *Schistosoma mansoni*. *Transactions of the Royal Society for Tropical Medicine and Hygiene* **65**: 432 (Laboratory demonstration)

HOCKLEY, D. J. and SMITHERS, S. R. (1970). Damage to adult *Schistosoma mansoni* after transfer to a hyperimmune monkey. *Parasitology* **61**: 95

KODANI, M. (1962). *In vitro* alteration of blood group phenotypes of human epithelial cells exposed to heterologous blood group substances. *Proceedings of the Society for Experimental Biology and Medicine* **109**: 252

KUSEL, J. R. (1970) Studies on the surfaces of cercariae and schistosomula of *Schistosoma mansoni*. *Parasitology* **61**: 127

LAHIRI, A. K., MITCHELL, W. M. and NAJJAR, V. A. (1970). The physiological role of the lymphoid system. VIII. The nature of the *in vitro* binding of erythrophilic γ-globulin and its effect on the configuration of the erythrocyte. *The Journal of Biological Chemistry* **245**: 3906

LEE, D. L. (1966). The structure and composition of the helminth cuticle. In *Advances in Parasitology* **4**: pp. 187. Ben Dawes (ed.). London and New York: Academic Press

MARCUS, D. M. and CASS, L. E. (1969). Glycosphingolipids with Lewis blood group activity: uptake by human erythrocytes. *Science* **164**: 553

MORRIS, G. P. (1971). The fine structure of the tegument and associated structures of the cercaria of *Schistosoma mansoni*. *Zeitschrift Parasitenkunde* **36**: 15

OLIVIER, L. J. (1966). Infectivity of *Schistosoma mansoni* cercariae. *American Journal of Tropical Medicine and Hygiene* **16**: 882

RAPPORT, M. M. and GRAF, L. (1969). Immunochemical reactions of lipids. *Progress in Allergy* **13**: 237. P. Kallos and B. H. Waksman (eds.). Basel and New York: S. Karger

ROBERTSON, J. D. (1964). Osmotic and ionic regulation. In *Physiology of mollusca*. Vol. I, 283. K. M. Wilbur and C. M. Yonge (eds.). London and New York: Academic Press

SCHENKEL-BRUNNER, H. and TUPPY, H. (1969). Enzymatic conversion of human O into A erythrocytes and of B into AB erythrocytes. *Nature (London)* **223**: 1272

SENFT, A. W. (1968). Studies in proline metabolism by *Schistosoma mansoni*. I. Radio-autography following *in vitro* exposure to radio-proline ^{14}C. *Comparative Biochemistry* **27**: 251

SMITH, J. H., REYNOLDS, E. S., and LICHTENBERG, F. VON (1969). The integument of *Schistosoma mansoni*. *American Journal of Tropical Medicine and Hygiene* **18**: 28

SMITHERS, S. R. and TERRY, R. J. (1965). Naturally acquired resistance to experimental infections of *Schistosoma mansoni* in the rhesus monkey (*Macaca mulatta*). *Parasitology* **55**: 701

SMITHERS, S. R. and TERRY, R. J. (1967). Resistance to experimental infection with *Schistosoma mansoni* in rhesus monkeys induced by transfer of adult worms. *Transactions of the Royal Society of Tropical Medicine and Hygiene* **61**: 517

SMITHERS, S. R. and TERRY, R. J. (1969). The Immunology of Schistosomiasis. *Advances in Parasitology* **7**: 41. Ben Dawes (ed.). London and New York: Academic Press

SMITHERS, S. R., TERRY, R. J. and HOCKLEY, D. J. (1969). Host antigens in Schistosomiasis. *Proceedings of the Royal Society B* **171**: 483

SPRENT, J. F. A. (1959). Parasitism immunity and evolution. In *The Evolution of Living Organisms*, p. 149. Victoria, Australia: Melbourne University Press

STIREWALT, M. A. (1963). Cercaria vs. schistosomule (*Schistosoma mansoni*): Absence of the pericercarial envelope *in vivo* and the early physiological and histological metamorphosis of the parasite. *Experimental Parasitology* **13**: 395

STIREWALT, M. A. (1966). Skin penetration mechanisms in helminths. In *Biology of Parasites*, p. 41. E. J. L. Soulsby (ed.). London and New York: Academic Press

SWEELEY, C. C. and DAWSON, G. (1969). Lipids of the erythrocyte. In *Red Cell Membrane: Structure and Function*. G. A. Jamieson and T. J. Greenwalt (eds.). Philadelphia and Toronto: J. B. Lippincott Co.

UHR, J. W. and MOLLER, G. (1968). Regulatory effect of antibody on the immune response. In *Advances in Immunology* **8**: 81. F. J. Dixon and H. G. Kunkel (eds.). New York and London: Academic Press

VOGEL, H. and MINNING, W. (1953). Uber die erworbene resistenz von *Macacus* rhesus gegenüber *Schistosoma japonicum. Zeitschrift für Tropenmedizin und Parasitologie* **4**: 418

CHANGES IN THE DIGESTIVE-ABSORPTIVE SURFACE OF CESTODES DURING LARVAL/ADULT DIFFERENTIATION

J. D. SMYTH

Department of Zoology and Applied Entomology,
Imperial College, London SW7 2BB

INTRODUCTION

Cestodes share with several other groups of Metazoa, notably the Acantho-cephala, the Trematoda, the Pogonophora, and—to a more limited extent—the entomophilic nematodes, the possession of a body covering which is not only protective, but is also a metabolically active surface through which nutrients can be absorbed and waste materials eliminated.

Although extensive work has been carried out on the cestode body covering, much of this has been concerned with structure rather than function. This is not surprising, for many functions can only adequately be studied in organisms isolated from host influences, i.e. by using *in vitro* systems—and very few such systems have been established at a suitable level of simplicity to make them readily usable. An exception to this restriction are those functional aspects which can be studied by short-term *in vitro* incubations, such as the uptake of carbohydrates or amino acids, and Read and his associates have been par-ticularly active in this field.

One of the possible ways in which clues to the functions of the various components of the body covering can be obtained is to examine their structure and metabolic behaviour during the various stages of the life cycle i.e as the parasite passes through biological environments which may have widely differ-ing characteristics. Yet surprisingly few such studies have been made, most workers being content with what might be termed 'static' studies i.e. the examination of the structure and/or physiology of one particular stage without necessarily relating these aspects to those pertaining in other stages of the life cycle. Some 'dynamic' studies, however, have recently been made and have contributed to our knowledge of the functional aspects of the cestode surface. In particular, evidence is accumulating that the activities of the cestode tegument —especially in relation to the way nutrients are obtained—are much more complex than formerly thought, and the body surface approaches what has been

41

called an 'absorptive-digestive' surface, on the basis of studies on the vertebrate alimentary canal (Crane, 1967, 1968, 1969). This concept involves a process known as 'membrane (= contact) digestion', evidence for the existence of which in cestodes was put forward in relation to *Echinococcus granulosus* (Smyth *et al.*, 1967) when it was first demonstrated that contact with a solid nutritive substrate 'triggered' differentiation of this species in a strobilar (rather than a cystic) direction.

As it is beyond the scope of this review to deal with all the functional aspects of the cestode surface, this paper deals with what is known regarding ultrastructural and histochemical changes during larval/adult differentiation in the two major Orders—the Pseudophyllidea and the Cyclophyllidea.

Before dealing with these aspects, it is necessary to briefly review the fine structure of the cestode tegument in general.

1. THE TEGUMENT: GENERALIZED STRUCTURE AND TERMINOLOGY

(a) GENERAL ACCOUNT

The extensive literature in this field has been reviewed by Lee (1966) and Smyth (1969). The essential features of the tegument are shown in Fig. 1. Various authors, however, have used different terminology for the contained structures and there is some confusion regarding interpretation. This terminology is reviewed below when dealing with the general features of the tegument.

The cestode tegument is essentially a syncytial epidermis made up of an outer anucleate cytoplasmic region, and an inner nucleated region containing 'cells' which are considered to represent sunken portions of the epidermis. Based on the terminology used for trematodes by Burton (1964) these regions have been termed the 'distal cytoplasm' and the 'perinuclear cytoplasm' respectively and they correspond approximately to the 'cuticular matrix' and 'subcuticular' cells of light microscopy. Other authors (e.g. Threadgold, 1965; Bråten, 1968a, b; Yamane, 1968, 1969) have used the terms 'external level' and 'internal level' for these regions. There appears to be no significant difference between these various terms but since there is some advantage in using the same terminology for both trematodes and cestodes, the terms distal and perinuclear have been used here.

(b) THE DISTAL CYTOPLASM

(i) 'Glycocalyx'

The external plasma membrane of the tegument (Fig. 1) is generally described as being covered with an 'amorphous' surface coat believed to be a mucopoly-

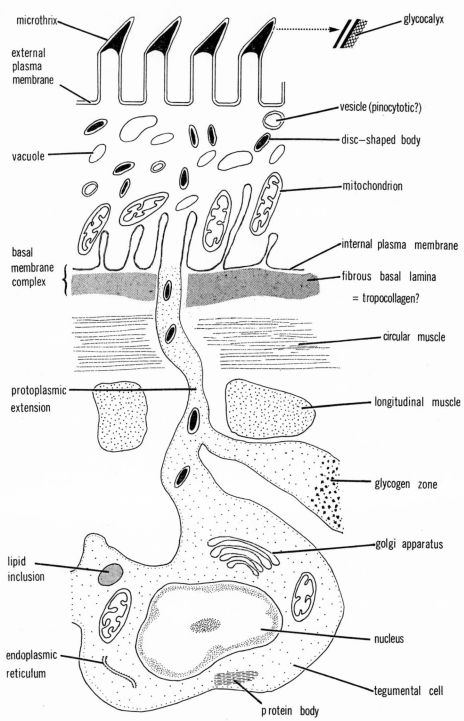

FIG. 1. Generalized diagram of the cestode tegument to illustrate the terminology used for various organelles.

saccharide or mucoprotein. Such layers are thought to be common to many if not all cell types (Bennett, 1963). This layer is commonly known as the 'glyco-calyx' (Morris and Finnegan, 1968) but it has also been referred to in various species as the 'filamentous coat' (Morris and Finnegan, 1969), or 'amorphous coat' (Lumsden, 1966a; Bråten, 1968a).

It should be noted, however, that Curtis (1967, 1972) believes that the histo-chemical and ultrastructural evidence on which the existence of this surface coat is based is equivocal and concludes that 'The evidence for the existence of a layer of material outside the plasmalemma is unsatisfactory at present'.

(ii) *External Plasma Membrane*

Some confusion also exists in the terminology used in describing biological membranes in general. This essentially arises in relation to the nature of the basic 'unit' membrane. It is generally accepted that a 'unit' membrane consists of two lipid layers with outwardly directed polar groups to which stabilizing layers of glycoproteins are attached, although Sjöstrand (1967) and Sjöstrand and Barajas (1968) disagree with this concept because frequently a globular structure is observed. Since at the electron microscope (EM) level a unit membrane appears as two electron dense lines with an electron lucid zone between, it is often referred to as a 'triple-layered' membrane. It is also some-times referred to as a 'double-layered' membrane if only the electron-dense layers are counted. Thus, Lumsden (1966a) refers to the external limiting membrane of *Lacistorhynchus tenuis* as being a 'trilaminate membrane 85–100 Å thick consisting of 2 parallel opaque components separated by a less dense intermediate layer 30–35 Å wide.' He later refers to this as 'the unit membrane'. Bråten (1968a) likewise describes the procercoid of *Diphyllobothrium latum* as being covered in a triple-layered membrane, and points out that the two dense layers making up this membrane are not the same thickness, the outer being 30 Å thick and the inner 55 Å with an electron lucid zone of 100–120 Å between. The limiting membrane of the plerocercoid of the same species has also been described by the same author (Bråten, 1968b) as being 'triple layered' (150 Å thick). Morris and Finnegan (1969) likewise refer to a 'trilaminate plasma membrane' in the plerocercoid of *Schistocephalus solidus*.

The limiting membrane of the tegument may prove to be more complex than formerly thought, for in the adult *Echinococcus granulosus* Jha and Smyth (1969), using magnifications of up to 190,000, have described the limiting membrane covering the microthrix shafts of this species as being made up of *two* unit membranes, closely opposed. The outermost of these (total thickness 115–145 Å) was made up of the usual two electron dense layers (each 40–55 Å) with a 40–60 Å gap. The inner (second) membrane which lay 40–60 Å below the outer membrane had a total thickness of 110–130 Å with two electron dense layers, each 40–55 Å thick with a 20–40 Å gap between the two. In this species, discontinuities or gaps in the membrane were described, the whole forming a peculiar mosaic pattern (rather like netting wire!). The electron-dense layers of these membranes in *Echinococcus* were further shown to have a globular substructure not unlike that described by Sjöstrand (1967) which, it was speculated, could be important in 'membrane' digestion (see later). Whether this apparent double unit-membrane structure is unique to *Echinococcus* (or is an artifact!) remains to be examined by other workers. In further support of a double structure being present, in some species at least, is the observation of Rosario (1962) that in *Hymenolepis* spp. the limiting membrane is double the thickness of a typical unit membrane.

(iii) *The Microtriches*

The surface projections of cestodes have been referred to as 'microvilli' by some workers (e.g. Lee, 1966) but the majority use the term 'microtriches' (singular *microthrix*).

Each microthrix consists of a pointed, electron-dense spine which is separated from the proximal less-dense region or base by a crescentic, membranous tube-like structure, or cap. The spines are generally considered to be made of a hardened material consistent with attachment to the host surface; the electron-dense spines appear to be absent in *Moniezia* (Howells, 1965).

In view of the length of cestodes, and the different host sites in which the scolex and the posterior region of the strobila might find themselves in the intestine, regional structural differences in size, form and concentrations of microtriches might be expected. Few studies on this aspect have, however, been carried out although differences between the microtriches in the scolex and proglottid regions in some species have been noted. Thus in *E. granulosus,* the microtriches in the scolex region are hooked, curved or even barbed—a structural feature which would greatly assist in maintaining close adhesion with the host tissue (Jha and Smyth, 1971). In this respect they differ from the normal type found in the strobilar region.

In *Diphyllobothrium erinacei*, Yamane (1968), too, noted differences between the relative lengths of the distal and proximal regions of the microtriches in the scolex and proglottids (Table 2). In the scolex, consistent with adaptation for attachment, the spine-like distal regions were longer than those in the proglottids, whereas, in the latter, the proximal regions were longer, consistent with their presumed absorbing function.

Several workers have described the shaft region of microtriches of several species as containing 'fibrils' (Béguin, 1966; Lumsden, 1966a; Bråten, 1968b). According to Jha and Smyth (1969) these structures probably represent microtubules—a structural interpretation which would agree with the reported occurrence of microtubules in the microvilli of the intestinal epithelium of mice (Mukerjee and Williams, 1967).

(iv) *Cytoplasmic inclusions: Terminology*

A number of different kinds of inclusions have been described in the distal and perinuclear cytoplasms but, since different authors have used different terms in describing these, it is not always possible to identify analagous structures unequivocally when comparing different species. The greatest confusion has probably arisen in relation to the rather ill-defined terms 'vacuoles' and 'vesicles', which have been widely used. On the other hand, inclusions such as mito-chondria—at least when fully formed with cristae—are easily identifiable. Mitochondrial precursors or partly-formed mitochondria are, however, more difficult to identify; this question is discussed further below when mito-chondrial biogenesis during differentiation is discussed (p. 56). The terms used to describe the various inclusions are discussed briefly below.

(v) *Vesicular inclusions*

All authors have noted the presence of vesicles of varying sizes in both the distal and perinuclear cytoplasms. Many descriptions of these are very brief and incomplete. It is difficult, for example, to know if the 'vesicles' described by Threadgold (1965) in *Proteocephalus* are analogous with the 'cuticular mem-brane bound vesicles' (60–125 µm in diameter) described in *Lacistorhynchus* by Lumsden (1966a), the 'vesicles' in *E. granulosus* described by Morseth (1967) or the 'vesicules' in *Caryophyllaeus laticeps* described by Béguin (1966). In the latter, also, numerous electron-dense, membrane-bound, rod-like structures have been described (Béguin, 1966); these probably correspond to the 'disc-like bodies' described in the plerocercoid and adult tegument of *Diphyllobothrium* (Bråten, 1968a), and in the plerocercoid of *Ligula* and *Schistocephalus* (Charles and Orr, 1968; Morris and Finnegan, 1969).

A number of authors have described evaginations of the limiting membrane which are suggestive of the formation of pinocytotic vesicles, for example, in *Dipylidium caninum* (Threadgold, 1962), *Proteocephalus pollanici* (Threadgold, 1965), *Echinococcus granulosus* (Jha and Smyth, 1969), *Lacistorhynchus tenuis* (Lumsden, 1966a), *Diplogonoporus grandis* (Yamane, 1969) and *Hymenolepis diminuta* (Rothman, 1963). Although Threadgold (1965) concluded, solely on morphological grounds, that some of the vesicles were 'pinocytotic vesicles', there appears to be no unequivocal experimental evidence that pinocytosis does occur at the cestode surface. The uptake of electron-dense particles, such as ferritin, is required to elucidate this point. Rothman (1967) appeared to show that this took place in *Hymenolepis diminuta* not by pinocytosis but by a process termed *transmembranosis*. Lumsden and his colleagues (1970) failed to confirm these results using ferritin, thorotrast or Pelikan ink.

(vi) *Pore canals*

In some species (e.g. *D. caninum*, Threadgold, 1962; *Hymenolepis diminuta*, Rothman, 1963; *D. grandis*, Yamane, 1969) the tegument appears to be pierced by canals, termed 'pore canals', possibly related to the so-called 'subcuticular canals' which, according to Yamane (1968), arise from invaginations of the basal membrane. In *D. grandis* these canals are reported as dividing into 2 branches and extending almost to the bases of the microtriches. Rothman (1963) suggested that the so-called pore canals may, in fact, be pinocytotic vesicles in formation. It is possible that some of the canal-like structures described in the tegument may represent the unusual pit-organelles (Fig. 9) described by Kwa (1970, 1972c) in *Spirometra erinacei*; these are discussed later (p. 63).

(vii) *Mitochondria*

All authors have described mitochondria in all species studied. As the structure, origin and function of these is of particular interest to the main topic of this paper—namely larval/adult differentiation—this topic is not dealt with here but is considered in detail later (p. 56).

(viii) *Basal membrane complex*

The innermost region of the distal cytoplasm, in most species studied, has been described as being bounded by a membranous structure (Fig. 1) referred to variously (in part or as a whole) as a 'basement membrane' (Lumsden, 1966a; Threadgold, 1962), a 'basement layer' (Bråten, 1968a), 'basal lamina' (Morris and Finnegan, 1969), 'basal membrane' (Charles and Orr, 1968; Jha and Smyth, 1971), or 'basement lamella' (Baron, 1971). This structure appears to be a unit membrane although its substructure has not been examined in detail.

Most workers have commented that this membrane is frequently thrown into folds or evaginations, the nature and function of which is a matter of some dispute. Threadgold and Read (1970) identified what they termed a 'multitubular complex' associated with the basal membrane in *H. diminuta*, although the precise region examined was not stated. This complex appeared to consist of rows of tubules orientated vertically or horizontally with respect to the basal membrane. These workers suggested that the complexes might function in water or ion transport or both. In a parallel series of observations on the same species, Reissig (1970) described a membranous structure consisting of 'arrays of sets' of lamellae randomly orientated with respect to the basal membrane;

he speculated that they might have a sensory function. These bodies were distributed throughout the surface of the proglottids. It is particularly interesting to note that Reissig (1970) sometimes found these membranes to be continuous with the basal membrane, for Jha and Smyth (1971) observed that in the scolex of *E. granulosus* the basal membrane apparently gave rise to membranous bodies (dumb-bell shaped bodies = umbomitochondria or protomitochondria?) and speculated that these bodies became transformed into mitochondria. Threadgold and Read (1970) commented that in *Hymenolepis* mitochondria were 'frequently associated with the apical or lateral margin of the complexes' —an observation consistent with the hypothesis of mitochondrial biogenesis outlined above for *Echinococcus*. This question is considered further on p. 56.

Beneath the basal membrane is found a thicker layer (200 Å in the plerocercoid of *D. latum*) referred to as a 'fibrous' or 'amorphous' layer. Morris and Finnegan (1969) point out that the fibrous component of the vertebrate basal membrane complex is thought to be tropocollagen, but there is no evidence regarding its composition in cestodes.

Although the membranes and fibrous components mentioned above appear to be readily identifiable as separate structures, the terms 'basement membrane', 'basal lamina' etc., have sometimes been used to describe the complex formed by the two components. In order to avoid confusion, it seems wisest to follow Morris and Finnegan (1969) and use the term 'basal' or 'basement membrane complex' (Fig. 1).

(c) THE PROXIMAL CYTOPLASM

As this Symposium is concerned largely with 'surface' phenomena, the structure of the distal cytoplasm is of particular interest, whereas that of the proximal cytoplasm is perhaps less so in this context. Nevertheless, since there is evidence of material being synthesised in the proximal cytoplasm and passed to the distal cytoplasm (see p. 65), the structure of the proximal cytoplasm must also be briefly outlined here.

There appears to be general agreement regarding the structure of the proximal cytoplasm, and only minor differences in the terminology of different authors can be noted. The proximal cytoplasm is connected to the distal cytoplasm by protoplasmic extensions which pass through the basement membrane complex and the muscular layers and lead into 'cells' which are part of the general epidermal syncytium. These appear to represent the sunken portions of the epidermis as in trematodes and turbellarians. It should be noted that Rothman (1963) referred to these connecting protoplasmic extensions as 'subcuticular canals'.

These 'cells' (Fig. 1) contain the usual cytoplasmic constituents: nucleus, mitochondria, golgi apparatus, endoplasmic reticulum and various special inclusions such as lipid and protein bodies and substantial amounts of glycogen; the latter substance is described as appearing in α and β form in some species.

2. FUNCTIONAL CHANGES IN THE TEGUMENT DURING LARVAL/ADULT DIFFERENTIATION

(a) GENERAL CONSIDERATIONS

Although, as outlined above, many workers have studied the ultrastructure of the tegument, little unequivocal evidence has been obtained regarding the function of the various contained components. It is known from extensive physiological studies (reviewed by Smyth, 1969) that, as in the vertebrate gut, various substances can be taken up by simple diffusion, mediated transport or active transport or combinations of these mechanisms. The presence of numerous vacuoles, vesicles or membrane-bound bodies of various sizes and densities in the tegument, as well as the presence of numerous metabolic enzymes, suggests that secretory (or excretory) processes of some kind are actively proceeding in this region. Although there has been much speculation on this, until recently, little experimental evidence has been available to provide clues regarding their possible function. One possible approach to this problem is to examine the ultrastructure of a larval form in its own environment and to examine the changes (if any) which occur when it passes into the environment provided by the definitive host and undergoes growth and differentiation to the adult form. If, at the same time, parallel studies are carried out on the biochemistry and cyto-chemistry of the differentiating organisms, one might expect to obtain some clues regarding the likely function of the contained organelles.

Yet, apart from the brief studies by Bråten (1968a, 1968b) and Yamane (1968) on the procercoid/plerocercoid/adult transformation in the Pseudo-phyllidea and those of Jha and Smyth (1971) on the hydatid cyst/adult transformation in *E. granulosus*, no studies of this kind appear to have been made. This work is reviewed below.

(b) PSEUDOPHYLLIDEA: DIFFERENTIATION IN *Diphyllobothrium* SPP.

(i) *The procercoid/plerocercoid/transformation*

In spite of the major differences in the environments encountered successively by this species during its life cycle, namely the copepod haemocoele, fish muscle and the mammalian intestine, the observed changes in the structure of the tegument are surprisingly small. Bråten (1968a, b) observed almost no

TABLE 1

Ultrastructural changes in distal cytoplasm of the tegument of *Diphyllobothrium latum* during procercoid/plerocercoid/adult differentiation; only major changes are noted (Data from Bråten, 1968a, b)

Procercoid (in copepod)	Plerocercoid (in Fish)	Adult worm (in hamster)	
		4-day (immature)	17-day (mature)
Amorphous coat (glycocalyx?)	+	+	+
External plasma membrane	150 Å	+	+
Microtriches	4–5 per (μm)^2 surface	?	25–30 per (μm)^2 surface; growth in size (see Table 2)
Disc-like bodies	+	+	+
Lamellated bodies	+	Greatly reduced in numbers	Absent
Mitochondria	+	?	Increase in size and number
Vesicles (pinocytotic?)	Absent	?	Absent

changes in ultrastructure between the procercoid and plerocercoid of *D. latum* apart from a slight thickening in the external plasma membrane in the plerocercoid and the absence of the membrane-bound vesicles (pinocytotic?) described in the procercoid (Table I). The functional significance of these changes are not clear because the functions of the membrane-bound vesicles are not known. The relatively minor nature of these changes may be related to the fact that both procercoid and plerocercoid occur in cold-blooded hosts in which the growth rate is relatively slow.

(ii) *The plerocercoid/adult transformation*

Differences at the ultra structural level of the distal cytoplasm were noted by Bråten (1968b) and Yamane (1968) when comparing the plerocercoids and adults of *D. latum* and *D. erinacei* respectively; some of these are summarized in Tables 1 and 2. Briefly, four major changes were noted in the plerocercoid/adult transformation:
1. A 6-fold increase in the number of microtriches per unit area (in *D. latum*).
2. A 4–7-fold increase in the length of the proximal part of the microtriches (in both species).

TABLE 2

Comparison of the size (in μm) of microtriches in the plerocercoids and adults of species of pseudophyllidean cestodes

Reference	Species	Microtriches		
		Distal part	Proximal part	
		Approx. length	Approx. length	Approx. width
Bråten (1968b)	*Diphyllobothrium latum*			
	Plerocercoid	1·0–1·5	0·2	—
	Adult	2·5	1·5	—
Yamane (1968)	*D. (Spirometra) erinacei*			
	Plerocercoid	2·0	0·4	0·3
	Adult			
	Scolex	3·5	0·4	0·4
	Proglottid	1·8	1·7	0·1

3.   An increase in the number of mitochondria and vesicles (in both species).
4.   The disappearance of one organelle in each case—the 'lamellated bodies' in *D. latum* and the 'disc-like' bodies in *D. erinacei*.

It can be speculated that the increase in the number of microtriches, the increase in the length of the (presumed?) absorptive region of the microthrix and the increase in the number of mitochondria can be related to the greater nutritive requirements and metabolic activities of the growing and differentiating adult worm. This result also provides indirect evidence of the importance of the microtriches in nutrition. With respect to the disappearance of the 'lamellated bodies' in *D. latum* and disc-like bodies in *D. erinacei*, it is tempting to conclude that this is related to the increase in mitochondria and other vesicles and that these bodies may play a part in the biogenesis of mitochondria—a process which has been shown to accompany larval/adult differentiation in the cyclophyllidean *E. granulosus* (Jha and Smyth, 1971) (see p. 56). On this view, the 'bodies' described in plerocercoids of *D. latum* and *D. erinacaei* may in fact represent 'protomitochondria' although no stages in the biogenesis of mitochondria, such as described in *E. granulosus*, have yet been observed in these pseudophyllidean species.

(c) CYCLOPHYLLIDEA: DIFFERENTIATION IN *Echinococcus granulosus*

(i) *The role of the surface in the induction of strobilization*

The larval stage (the protoscolex) of the hydatid organism, *Echinococcus granulosus* offers unusually interesting material for studies on morphogenesis and differentiation for it possesses the unusual potential of differentiating in one of two directions depending on its location in the host. Thus, when a protoscolex in a hydatid cyst is eaten by a dog and passes to the gut, it differentiates into a strobilated tapeworm. On the other hand, if a protoscolex is injected into the body cavity of a vertebrate host—or ruptures from a cyst into the body cavity in its intermediate host—it dedifferentiates in a cystic direction, eventually forming another hydatid cyst with brood capsules and protoscoleces.

Experiments *in vitro* have shown that when a suitable gas phase, nutrient medium etc. are provided, the factor initiating differentiation in a strobilar direction appears to be the contact of the scolex surface with a suitable solid, nutritive substrate (Smyth, 1967; Smyth *et al.*, 1967). The most suitable substrate found has been serum coagulated at 70–80°C; bovine serum being superior to horse or dog serum. Other protein-containing bases such as gelatine or collagen were unsuccessful in inducing strobilization and non-nutrient bases such as agar were also unsuccessful. The precise experimental conditions under which this induction of strobilization occurs *in vitro* have been reviewed elsewhere (Smyth, 1969). This result indicated that the scolex plays a very important role in control of differentiation in this species, and it was considered that a study of the ultrastructure of this region, before and during the early stages of strobilar differentiation, might yield information regarding the nature of the controlling mechanism.

(ii) *Possible nature of the 'contact' stimulus*

Before dealing with the ultrastructure of this region let us speculate regarding the possible nature of the 'contact' stimulus inducing differentiation in a strobilar direction. A number of possibilities suggest themselves: thus the stimulus could be:

1. Nervous in origin, i.e. a tactile response from a suitable substrate.
2. *Directly* nutritional in origin, i.e. the substrate provides a *specific* substance or substances which either diffuses out of the substrate and is absorbed, or ingested in solid form by pinocytosis and digested intracellularly.
3. *Indirectly* nutritional in origin via *extracellular digestion* i.e. the scolex releases an enzyme which digests the substrate, and the by-products of this digestive process are absorbed and initiate strobilization.
4. *Indirectly* nutritional in origin via *membrane digestion* i.e. the scolex surface contains membrane-bound enzymes which, on contact with a suitable substrate, digest it and the byproducts are immediately absorbed and initiate strobilization.

It is thought that none of hypotheses (1)–(3) are viable and that (4) i.e. membrane digestion, may be operating in this case. The reasons for this tentative conclusion have been discussed in detail elsewhere (Smyth, 1969), but are outlined briefly below. It must be borne in mind, however, that these hypotheses are not mutually exclusive since, for example, absorbed materials may have both a nutritional and a stimulatory effect.

That the stimulus is tactile in origin is rejected on the grounds that a surface such as agar with a similar consistency does not induce strobilization, although this conclusion is not necessarily unequivocal since a chemotactic mechanism could also be involved.

That the stimulus does not appear to be a result of the uptake of a substance diffusing out of the substrate, is suggested by the fact that if the protoscoleces are confined within a semipermeable cellulose tube, when grown in a diphasic culture (i.e. in a situation where they would have an opportunity to take up any small molecular material diffusing from the solid base) strobilization does not occur. Hence, as far as present evidence goes, actual *contact* of a protoscolex with the substrate appears to be necessary. Direct pinocytosis is, at present, rejected on the grounds that pinocytosis has not yet been demonstrated in cestodes. Preliminary experiments, by the writer, using ferritin also failed to demonstrate uptake by pinocytosis in *Echinococcus*.

Rejection of the above possibilities led to the hypothesis that the protoscolex surface was capable of digesting the substrate surface and that the by-products of digestion were absorbed and form the basis of the strobilization stimulus. Yet repeated attempts (Shields and Smyth, unpublished work) to demonstrate the release of enzymes from the surface of developing scolices, utilizing fresh frozen sections and the photographic plate technique of Adams and Tuqan (1961) have failed to demonstrate extracellular enzyme activity of any kind. It was thus concluded that, if a digestive process took place in this region, the enzymes were not released into the medium but were surface- or membrane-bound; i.e. 'membrane digestion' operated at the host-parasite interface.

If this were so, it could be expected that if the interface was constantly disturbed by the culture medium being made to flow rapidly instead of being static (or almost so), the by-products of this digestive process would be swept away from the enzyme site and sufficient could not be absorbed by the scolex to induce strobilization.

An *in vitro* experiment was therefore set up in which *in the same medium* worms were cultured under conditions under which the interface was 'undisturbed' (cultures B_1, B_2 and control) and 'disturbed' (cultures A_1 and A_2). The apparatus utilized is shown in Fig. 2, in which the liquid phase is circulated by means of a gas 'lift'; technical details of the culture medium, solid phase etc. are given by Smyth (1967). It can be seen that the medium in the upper chamber is in continual circulation, i.e. is 'disturbed', whereas that in the lower chamber is initially 'undisturbed', although exchange of materials with those circulating through the cellulose tube is possible by diffusion.

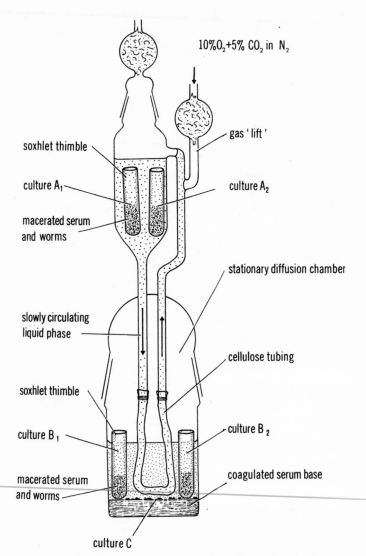

FIG. 2. Circulatory apparatus to compare the effect of culturing evaginated protoscoleces of *Echinococcus granulosus* under 'disturbed' conditions (Cultures A_1 and A_2) and 'undisturbed' conditions (Cultures B_1 and B_2 and Control culture C). A_1 and A_2, B_1 and B_2 are soxhlet thimbles containing macerated (previously coagulated) serum in which worms are embedded. In C, the worms lie on the surface or in holes on a solid coagulated serum base. In 'disturbed' conditions, the medium is flowing continually through the culture; in 'undisturbed' conditions, circulation is confined to a cellulose tubing through which exchange by diffusion can take place, but the parasite-substrate interface is relatively undisturbed. Results are shown in Fig. 3. Segmentation and sexual maturity occurred only under 'undisturbed' conditions (i.e. in cultures B_1, B_2 and C). Not to scale.

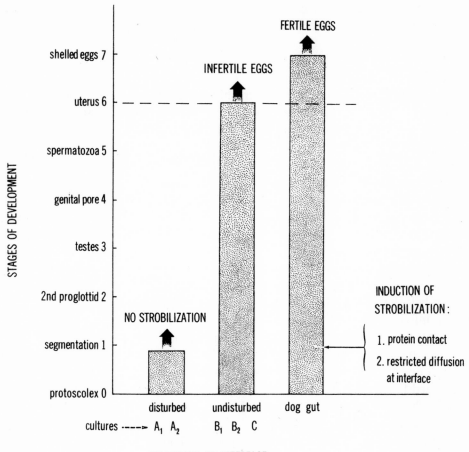

FIG. 3. Differentiation of *Echinococcus granulosus* under 'disturbed' and 'undisturbed' conditions *in vitro*, using the apparatus shown in Fig. 2. Disturbing the parasite-substrate interface (Cultures A_1 and A_2) greatly inhibits development. Under 'undisturbed' conditions (Cultures B_1, B_2 and C), development approaches that in the dog, although fertilization does not occur *in vitro*.

The lower chamber contained a solid substrate (coagulated bovine serum) on which culture **C** was placed as a control, and both upper and lower chambers contained soxhlet thimbles containing macerated coagulated bovine serum containing cultures A_1, A_2 and B_1, B_2 respectively. The whole was incubated at 38·9°C (dog body temperature).

Results are shown in Fig. 3. Only those worms in the undisturbed conditions (cultures series **B** and **C**) underwent strobilization and, subsequent differentiation of genitalia. This result, while not demonstrating unequivocally that membrane digestion takes place, is, at least, consistent with this hypothesis. It would also appear to support the view that absorption of diffusible substances from the coagulated serum does not provide the stimulus because such substances would clearly be available to both 'disturbed' and 'undisturbed' cultures.

(iii) *Ultrastructural changes during protoscolex/adult differentiation*

Before considering the concept of membrane digestion and its possible significance in *Echinococcus* in particular, and cestodes in general, the changes which occur during the differentiation of the protoscolex to a strobilating tapeworm must first be considered.

Jha and Smyth (1971) have shown that in *E. granulosus*, during the very early stages of strobilization, marked changes occur in the composition of the cytoplasmic constituents; moreover, these changes appear to be correlated with changes in activity of a small group of gland cells in the tip of the scolex forming the so-called rostellar gland (Smyth, Morseth and Smyth, 1969). Briefly, in a freshly evaginated protoscolex (i.e. one removed from a hydatid cyst and evaginated with bile under suitable physiological conditions) the distal cytoplasm of the rostellum contains very few mitochondria but is rich in unusual dumb-bell shaped membranous bodies (Fig. 4). The latter appear to be formed by an infolding of the basal membrane.

After 24–48 hours *in vitro* culture, i.e. at the early stages of differentiation in a strobilar direction, the dumb-bell shaped bodies appeared to be evaginating to form loops and differentiating into mitochondria. Newly formed mitochondria possessed few cristae. In well established worms, the distal cytoplasm was rich in mitochondria some of which were elongated and some with profiles arranged in a concentric manner.

Perhaps the most unexpected finding by Jha and Smyth (1971) in the scolex of *Echinococcus* was that 'free' mitochondria could be detected in the secretion droplets formed at the tip of the rostellum and were also found adhering to the outside surface. These authors suggested that these 'free' mitochondria originated from what appeared to be membrane-bound 'mitochondrial sacs'. Although some of the latter may prove to be lysosomes—a view not inconsistent with the role of exocytosis of mitochondria at the surface—further examination has revealed (as suggested during the Symposium Discussion) that some of the 'sacs' figured are, in fact, unusual sections across sensory endings; these similarly are sac-like and contain numerous mitochondria. Thus, although mitochondria are found on the outside surface of *Echinococcus*, their actual method of transport to the outside is still open to question and requires further critical examination.

Jha and Smyth (1971) also described the presence of 'lamellar bodies' each of which was surrounded by several membranes in a concentric manner, the core containing tightly packed membrane-bound vesicles. No functional significance could be attributed to these.

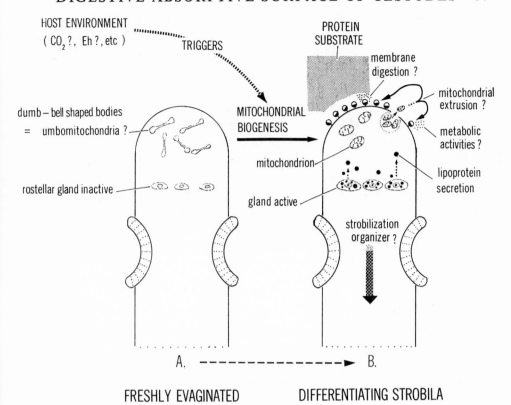

HOST ENVIRONMENT
(CO_2?, Eh?, etc)   TRIGGERS

PROTEIN
SUBSTRATE

membrane
digestion ?

mitochondrial
extrusion ?

dumb – bell shaped bodies
= umbomitochondria ?

MITOCHONDRIAL
BIOGENESIS

metabolic
activities ?

mitochondrion

lipoprotein
secretion

rostellar gland inactive

gland active

strobilization
organizer ?

A. ----------------▶ B.

FRESHLY EVAGINATED
PROTOSCOLEX

DIFFERENTIATING STROBILA

FIG. 4. A highly speculative hypothesis concerning the functional changes in the rostellar region of *Echinococcus granulosus* during early strobilar differentiation. It is speculated that when the scolex enters the gut environment (1) mitochondrial biogenesis is triggered, (2) the mitochondria appear to develop from pre-existing dumb-bell shaped bodies (umbomitochondria? protomitochondria?) with possibly a contribution from the rostellar gland, (3) some mitochondria are extruded to the outside with the rostellar secretion, (4) extruded mitochondria either (a) engage directly in metabolic activities or (b) disintegrate and release their enzymes which attach to the glycocalyx and take part in digestive and/or metabolic activities.

The rostellar region also contains, just below the perinuclear cytoplasm, a group of cells comprising the 'rostellar gland' (Smyth, 1964). Smyth, Morseth and Smyth (1969) have shown that during the protoscolex/strobila transformation the cells of this gland, which are inactive in the protoscolex, become increasingly active and secrete globules which pass into the distal cytoplasm (Fig. 4, B). Moreover, ultrastructural studies showed that this secretion had its origin in the nucleus and, furthermore, was lipoprotein in nature. Jha and Smyth (1971) have speculated that the secretions of the rostellar gland (Fig. 4, B) may contribute to biogenesis of mitochondria, which is proceeding actively in the distal cytoplasm, a hypothesis in keeping with the widely held view that mitochondria have a dual origin. Its association with the nucleus thus may be of special significance, for it is known that mitochondria contain DNA and the process described here could be compared with mitochondrial biogenesis in the haemoflagellates where the mitochondria has its origin in the DNA-containing kinetoplast.

In the haemoflagellates and also in malarial parasites, there is much evidence (reviewed by Peters, 1970) that mitochondrial biogenesis is associated with a 'switch' in metabolism during passage from one host to another.

The apparent onset of mitochondrial biogenesis when *Echinococcus* passes from the intermediate (sheep) host to the definitive (dog) host—or the equivalent *in vitro* stage—would appear to point to the likely occurrence of a similar metabolic switch in this organism also. We have no experimental evidence, however, that such a switch does occur, but this question is under investigation by histochemical and biochemical means. If this can be demonstrated, it would reveal a uniformity of metabolic behaviour, at present not apparent, between prokaryote and eukaryote parasites.

3. THE CESTODE TEGUMENT AS A DIGESTIVE-ABSORPTION SURFACE

(a) THE VARIOUS DIGESTIVE PROCESSES

Before considering the significance of the ultrastructural changes described above in relation to the possibility that membrane digestion is taking place, a brief account of this latter process as put forward by Ugolev (1965, 1968)—but which has not received much attention by biologists—is given below:

The differences between extracellular, intracellular and membrane digestion are illustrated in Fig. 5.

(i) *Extracellular digestion* (Fig. 5A).

This is characterized by the fact that the enzymes, synthesised by the cell are passed into the extracellular medium and hence act some distance away from the secreting cell. Ugolev (1968) refers to digestion which takes place in a special region or cavity as 'cavital' digestion.

(ii) *Intracellular digestion* (Fig. 5B)

This is used for processes whereby intact or partly broken down food is taken into the cell and there undergoes further hydrolysis. Intracellular digestion can take place in special digestive vacuoles formed during phagocytosis and pinocytosis.

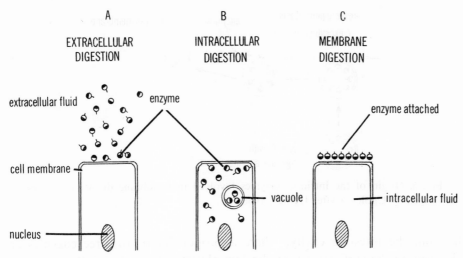

FIG. 5. Localization of enzyme hydrolysis during (A) extracellular, (B) intracellular and (C) membrane digestion (based on Ugolev, 1968).

(iii) *Membrane* (= *contact*) *digestion* (Fig. 5C)

It is beyond the scope of this paper to deal with the detailed experimental evidence that such a process as membrane or contact digestion does occur in biological systems. Ugolev (1968) has reviewed this evidence in detail and the following summarizes the chief characteristics of this process: (Terms utilized by Ugolev are in parentheses).

1.   The enzymes involved in the digestive process are fixed on the cell membrane.

2.   The process of digestion occurs on the membranes which divide the intracellular and extracellular environments.

3.   'Effective interaction between the processes of hydrolysis and transmembranal shift takes place on the cell membrane surface.'

4.   The enzymes are adsorbed at 'definite points on the membrane' which can be considered to be 'specific absorption centres'.

5.   'It is established that disaccharides, peptidases, as well as, apparently, nucleotidases, alkaline phosphatase, and monoglyceridase, are firmly bound to membrane structures and do not pass into the cavity of the small intestine.' Other 'intestine-proper' enzymes—pancreatic amylase, proteases, and lipase are distributed 'between the cell surface and the surrounding fluid'.

6.   The enzymes adsorbed on the surface are believed to originate from two sources:

(a) from within the cell, passing to the surface from intracellular vacuoles by a process of exocytosis (Ugolev's 'reverse pinocytosis', Fig. 6) to be adsorbed on the surface.

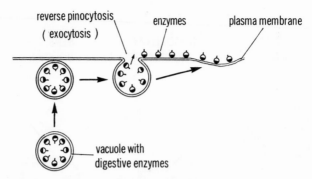

FIG. 6. Origin of the intrinsic enzymes involved in membrane digestion; as envisaged by Ugolev. (After Ugolev, 1968).

(b) from the intestine cavity, where a proportion of the free enzymes are adsorbed on the surface; 'an equilibrium always exists between the enzymatic activity of the intestinal content and the intestinal epithelium with respect to these enzymes' (see 5 above).

The evidence that membrane digestion occurs in the vertebrate intestine has led to the concept of a digestive-absorptive (Fig. 7) surface by Crane (1967, 1968, 1969) i.e. one which is both involved in digesting and absorbing and there is increasing histochemical, biochemical and physiological evidence in support of the vertebrate intestinal mucosa as being a surface of this nature.

(b) MEMBRANE-DIGESTION IN *Echinococcus granulosus*: ORIGIN OF MEMBRANE-BOUND ENZYMES

The ultrastructural changes described in the scolex during the protoscolex/ adult transformation of *Echinococcus*, in addition to fitting in with the general concept of a digestive-absorptive surface, would appear to throw some further light on the process of membrane digestion itself.

One of the fundamental questions in considering membrane digestion is the origin of the adsorbed enzymes. As indicated above, Ugolev (1968) distinguished between those of intrinsic origin (i.e. produced within the cytoplasm and passed by exocytosis to the surface) and those of extrinsic origin (i.e. absorbed from the intestinal lumen); Crane (1968) has commented that Ugolev has not always distinguished between intrinsic and extrinsic enzymes.

In *Echinococcus*, the fact that strobilar development has been induced *in vitro* only on contact with a solid (protein-containing?) substrate and that this phenomenon is accompanied by (a) a massive increase in mitochondrial biogenesis, and (b) the occurrence of 'free' mitochondria in the rostellum 'secretion' and at the surface, leads to the speculation that these enzymes may have their

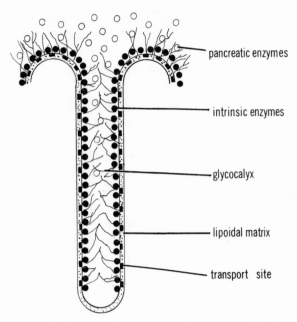

FIG. 7. The concept of a digestive-absorptive surface as exemplified by the vertebrate intestinal mucosa; based on Crane (1968); some details omitted.

origin in the released mitochondria, and/or from the sacs (= lysosomes?) which may transport them to the glycocalyx region of the tegument. Such an unusual role for mitochondria, although clearly out of line with the respiratory activities of (say) vertebrate mitochondria, is in keeping with the concept of a mitochondrion as an enzyme-containing sac. It is important, however, to emphasize that to a large extent our concept of the role of mitochondria has been dominated by work on those most used as biochemical models, such as rat liver or beef heart mitochondria. Yet, as Weinbach and von Brand (1970) have stressed, the role of mitochondria in essentially anaerobic parasites, such as cestodes, presents an intriguing paradox. They have further shown that the mitochondria of larval and adult *Taenia taeniaeformis*, although capable of oxidizing α-glycerophosphate and therefore probably capable of performing an α-glycerophosphate (Bücher) cycle, are incapable of oxidative phosphorylation which is generally considered the hallmark of mammalian mitochondria. It is also now recognized that mitochondria contain DNA and are capable of mutation; moreover, within the same organism or even within the same cell they may show heterogeneity (Wagner, 1969). Hence, there appears to be no basic reason why mitochondria could not become adapted to producing digestive enzymes, such as peptidases, capable of taking part in membrane digestion, although no direct experimental evidence can be provided in support of such a hypothesis.

It is therefore of interest to note that in the mucosa of the small intestine of the rat, Robinson (1963) has found both dipeptidase and alkaline phosphatase activity associated with the mitochondrial fraction of the cell. Rademaker and Soons (1957) have also shown that the proteolytic systems of the liver are localized largely in the mitochondria.

Additional evidence in support of a possible mitochondrial association with membrane digestion comes from ultrastructural studies of the vertebrate gut. Johnson (1967) described 60 Å knobs attached to the luminal surface of the plasma membrane of the hamster intestine brush border. As Crane (1968) points out, their morphology thus closely resembles the enzymatically active particles of inner mitochondrial membranes. Similar particles have been identified in cestode mitochondria (Harlow and Bryam, 1971). The presence of bile salts in the gut could assist the breakdown of the lipoprotein membranes of released mitochondria and, indeed, surface-active agents—such as the bile salt, sodium deoxycholate—are used extensively in laboratory work for rupturing mitochondria.

The hypothesis that released mitochondria or their contained enzymes may be involved in membrane digestion in the scolex region can only be regarded as highly speculative at present. Such a view does not necessarily exclude the possibility that other organelles such as vesicles and vacuoles, which may contain enzymes, could also take part in such a process, both in the scolex and in the strobila.

Jha and Smyth (1971) proposed an alternative hypothesis to that outlined above, namely, that the released mitochondria could play a metabolic rather than a digestive role. Presumably, mitochondria passed to the surface could make contact with substrates more rapidly and perhaps more efficiently, than within the cell, for substrate penetration problems would be avoided by this mechanism. On this view, such mitochondrial activity would result in an increased or altered metabolism which could, directly or indirectly, act as a strobilization trigger or stimulus. Although cestode mitochondria may not, in some species (e.g. *Taenia taeniaeformis*) be capable of oxidative phosphorylation, as mentioned above, they show α-glycerophosphate oxidase activity suggesting participation in an α-glycerophosphate (Bücher) cycle. Since glycerol is likely to be freely available in the vertebrate gut, released mitochondria would find an important substrate for their activities readily available. Again, mitochondria are believed to be involved in carbon dioxide fixation and, since this gas is readily available, released mitochondria could also engage in this activity. The hypothesis that extruded mitochondria, or their released or contained enzymes, act in a metabolic capacity is perhaps more probable, on theoretical grounds, than the view that they are engaged in digestive activities.

Although 'extrusion' of mitochondria appears to be a most unusual phenomenon, several other cases have been described. Thus, in the walls of the alveoli

in mammalian lungs, mitochondria extrusion takes place, the discharged mitochondria being apparently responsible for secreting a surface active fluid
(Clements, 1962). The process has also been reported in reticulocytes of dogs
which had been made anaemic by treatment with phenylhydrazine (Simpson
and Kling, 1968) and 'free' mitochondria have been observed in the gland
secretions of the adhesive organ of the trematode *Cyathocotyle bushiensis*
(Erasmus and Ohman, 1965).

The phenomenon of mitochondria extrusion in *Echinococcus* appears to be
associated with differentiation in a strobilar direction. To what extent it is

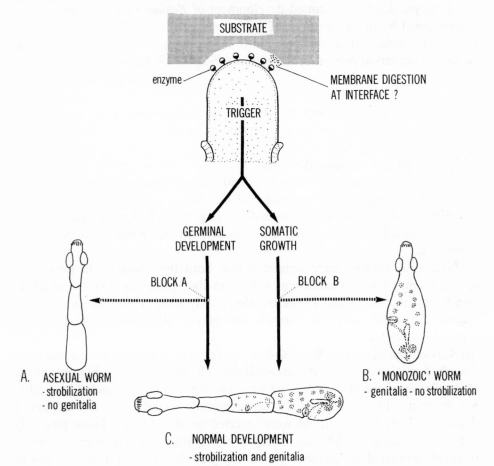

FIG. 8. Differentiation in *Echinococcus granulosus in vitro*. Germinal development
and somatic growth—although normally closely coordinated—appear to be able
to operate independently for (A) 'asexual' worms (i.e. segmented but without
genitalia and (B) 'monozoic' worms (i.e. with one set of genitalia, but not divided
into proglottids), in addition to (C) normal worms (with proglottids and genitalia)
have been formed *in vitro*. It is speculated that metabolic 'blocks' may inhibit either
developmental pathway. Based on Smyth (1971).

concerned with growth as well as differentiation cannot be stated at this stage. The process of somatic growth and genital differentiation appear to be able to operate independently of each other in this species for, under certain conditions *in vitro,* 'monozoic' forms (i.e. those with genitalia but unsegmented; Fig. 8B) can develop (Smyth, 1971). Conversely, segmented forms without genitalia have also been produced *in vitro*, (Fig. 8A). It is speculated that this interference with the 'normal' developmental pattern is due to inhibition of uptake of essential materials, perhaps as a result of the development of 'immune' precipitates or other abnormal conditions at the parasite/medium interface. The evidence for this is as yet slight, but stunted development of *Echinococcus in vitro* is often accompanied by the appearance of 'secretions' at the rostellar tip or in other parts of the tegument, a phenomenon itself probably associated with immune factors in the serum present in the medium (Smyth, 1969).

(c) EVIDENCE OF MEMBRANE DIGESTION IN OTHER CESTODE SPECIES

Although the phenomena described above in *Echinococcus* have not been described in other cestodes, there is considerable indirect evidence from other species which supports the occurrence of membrane digestion and the concept of a digestive-absorptive surface in cestodes in general. This additional evidence is histochemical, ultrastructural, and biochemical but much of it is, as yet, equivocal.

Numerous authors have suggested that both the cestode and trematode teguments may have a secretory as well as an absorptive function, but in addition to Smyth, Miller and Howkins (1967), only Kwa (1970) appears to have suggested that these secreted materials may play a role in membrane digestion. Kwa (1970, 1972c) described two universal organelles, 'pit organelles' and 'packets of granules' (Fig. 9) in *Spirometra erinacei*, neither of which had been described previously in the Pseudophyllidea. The rarer of these (Fig. 9), the pit organelle, consisted of a sunken pit whose entrance was surrounded by a cilia-like structure. The pit contained large membrane-bound bodies 0·2–0·3 µm in diameter which Kwa believed were secreted to the outside. These pits only appeared to occur at the anterior tip of the sparganum. The second type of organelle consisted of 'packets' containing granules (Fig. 9) 0·06–0·08 µm in diameter which may possess a membranous substructure. Similar membrane-bound granules were seen on the outer surface of the tegument and it was suggested that these granules are synthesized in the proximal cytoplasm and passed to the distal cytoplasm where they are finally secreted and may take part in membrane digestion. In addition, Kwa (1972b) demonstrated that the scolex of *S. erinacei* possessed proteolytic activity but whether the enzyme responsible

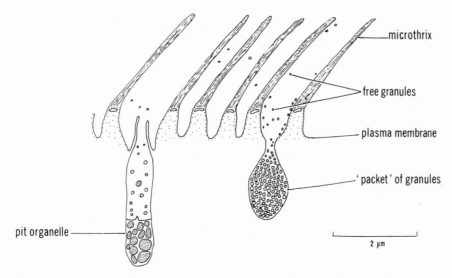

FIG. 9. Two types of secretory tegumental organelles described in *Spirometra erinacei* by Kwa (1970, 1972c). (Courtesy Mr B. H. Kwa; diagram considerably simplified from his original figures.)

was secreted or bound to the surface was not determined; no 'penetration' glands were detected (Kwa, 1972a).

Although organelles similar to those in *S. erinacei* have not been observed in other cestodes, the small area covered by electron microscope sections may have resulted in these being overlooked. Several authors have noted the presence of finger-like 'projections' of the distal cytoplasm (Timofeev, 1964; Threadgold, 1965; Charles and Orr, 1968) and these may well represent special secreting structures. It is also possible that the various rod-shaped bodies, vacuoles or vesicles may represent stages in a secretory process although, if this were so, it is difficult to understand how it is that the actual process of secretion has not been observed in spite of the numerous studies made on many species.

There is also biochemical and physiological evidence to support the concept of membrane digestion in cestodes. Thus, Lumsden (1966b), on the basis of isotope studies with tritiated proline on *H. diminuta*, showed that a significant percentage of the radioactivity was incorporated into the proximal cytoplasm and suggested that 'structural and non-mitochondrial' enzymes were synthesized in this region and then transported and secreted into the distal cytoplasm. It would be reasonable to assume that some of this material would be further secreted to the exterior.

On chemical grounds, Taylor and Thomas (1968) provided evidence for the occurrence of membrane digestion in *Moniezia expansa*. They found that the presence of the living tapeworm increased the rate of hydrolysis of starch by α-amylase and, moreover, the degree of acceleration was directly proportional to the surface area of the worm although it varied with pieces from different regions. This result (which requires confirmation under sterile conditions to be unequivocal) could account for an interesting observation by Read and Rothman (1957) that *H. diminuta* grown in rats fed on starch, grew twice as fast as those fed on glucose. As Taylor and Thomas (1968) point out, if membrane digestion was occurring on the surface of both host and parasite, the byproducts of hydrolysis would be available to both host and parasite. In contrast, in rats fed on glucose, the latter would probably be more readily absorbed by the host and not so easily available to the parasite.

Arme and Read (1970) showed that *H. diminuta* was virtually impermeable to ^{14}C-fructose diphosphate but that the intact worm hydrolysised this substance and released fructose into the medium. These workers concluded that this indicated that hydrolysis of the hexose diphosphate was occurring at the surface of the worm. It is questionable, however, if this result is unequivocal, for the enzyme may have been secreted into the medium and hydrolysis may not necessarily have taken place at the surface.

There is, however, supporting cytochemical evidence that the hydrolysis of phosphate esters takes place at the surface for Rothman (1966) in *Hymenolepis citelli* and Lumsden and colleagues (1968) and Dike and Read (1971) using ultrastructural cytochemistry showed that phosphohydrolysases were localized at or on the external surface in this species.

On the basis of preliminary experiments using fractions containing largely tegument, Bailey and Fairbairn (1968) concluded that in *H. diminuta* mono-olein was hydrolysised by a lipase at or near the surface.

4. MEMBRANE DIGESTION AND ACQUISITION OF 'HOST-LIKE' ANTIGENS

It is beyond the scope of this paper to discuss the extensive work which has been carried out by groups such as Smithers and his co-workers in England (e.g. Smithers *et al.*, 1969; Clegg *et al.*, 1971) and Capron and his co-workers in France (e.g. Capron *et al.*, 1968) on the apparent possession of 'host-like' antigens by trematodes, cestodes and nematodes. Nevertheless, it is obvious that if cestodes—and probably trematodes—are capable of attaching host enzymes to their outer surfaces (or in the case of nematodes to their intestinal brush border) this would result in a parasite surface coated in host antigens. Furthermore, the situation could be complicated by the addition of secreted enzymes of parasite origin. How relevant this concept is to the position of schistosomes which live in a (theoretically) non-digestive environment, it is not possible to say, but it can be speculated that even here, the trematode cuticle may be behaving as a digestive-absorptive surface. Certainly, in the case of cestodes, the process of membrane digestion would appear to confer an additional benefit on the parasite namely, by coating it with host enzymes, to make it antigenically more acceptable to the host. Considerable work is, however, clearly needed in relation to this hypothesis.

CONCLUSIONS

The evidence outlined above points to the conclusion that the cestode integument is an organ of considerable structural and functional complexity which has properties in keeping with the concept of a digestive-absorptive surface. In addition to its interest as a parasitological problem, consideration of this concept raises fundamental questions of great biological importance such as mitochondrial biogenesis, enzyme synthesis and release, antigenic protection, and cell and tissue differentiation; some of these questions have been dealt with here. At this stage in our knowledge, many of the hypotheses put forward concerning the larval/adult transformation of cestodes and the possible relationships between mitochondrial activities, membrane digestion and differentiation —in *Echinococcus* in particular—must be regarded as highly speculative. Considerable further work at tissue, cellular, ultrastructural and biochemical levels will clearly be needed if further progress—in what appears to be an exciting field—is to be made.

ACKNOWLEDGEMENTS

Some of the work presented here was supported by grants from the Australian Meat Research Board to which grateful acknowledgement is made.

My thanks are also due to Mr B. H. Kwa for permission to utilize data and a text-figure (Fig. 9) from some of his work *in press*.

REFERENCES

ADAMS, C. W. M. and TUQAN, N. A. (1961). The histochemical demonstration of protease by a gelatine-silver film substrate. *Journal of Histochemistry and Cytochemistry* 9: 469–72

ARME, C. and READ, C. P. (1970). A surface enzyme in *Hymenolepis diminuta* (Cestoda). *Journal of Parasitology* 56: 514–16

BAILEY, H. H. and FAIRBAIRN, D. (1968). Lipid metabolism in helminth species. V. Absorption of fatty acids and monoglycerides from micellar solution by *Hymenolepis diminuta* (Cestoda). *Comparative Biochemistry and Physiology* 26: 819–36

BARON, P. J. (1971). On the histology, histochemistry and ultrastructure of *Raillietina cesticillus* (Molin, 1858). Fuhrmann, 1920 (Cestoda, Cyclophyllidea). *Parasitology* 62: 233–45

BÉGUIN, F. (1966). Étude au microscope électronique de la cuticle et de ses structures associées chez quelques cestodes. Essai d'histologie comparée. *Zeitschrift für Zellforschung und Microskopische Anatomie* **72**: 30–46

BENNETT, H. S. (1963). Morphological aspects of extracellular polysaccharides. *Journal of Histochemistry and Cytochemistry* **11**: 14–23

BRÅTEN, T. (1968a). An electron microscope study of the tegument and associated structures of the procercoid of *Diphyllobothrium latum* (L.). *Zeitschrift für Parasitenkunde* **30**: 95–103

BRÅTEN, T. (1968b). The fine structure of the tegument of *Diphyllobothrium latum* (L.). *Zeitschrift für Parasitenkunde* **30**: 104–12

BURTON, P. R. (1964). The ultrastructure of the integument of the frog lung-fluke, *Haematoloechus medioplexus* (Trematoda: Plagiorchiidae). *Journal of Morphology* **115**: 305–18

CAPRON, A., BIGUET, J., VERNES, A. and AFCHAIN, D. (1968). Structure antigénique des helminthes. Aspects immunologiques des relations hôte-parasite. *Pathologie et biologie. Paris* **16**: 121–38

CHARLES, G. H. and ORR, T. S. C. (1968). Comparative fine structure of outer tegument of *Ligula intestinalis* and *Schistocephalus solidus. Experimental Parasitology* **22**: 137–49

CLEGG, J. A., SMITHERS, S. R. and TERRY, R. J. (1971). Concomitant immunity and host antigens associated with schistosomiasis. *International Journal for Parasitology* **1**: 43–9

CLEMENTS, J. A. (1962). Surface tension in the lungs. *Scientific American* **207**: 120–30

CRANE, R. K. (1967). Structural and functional organization of an epithelial cell brush border. In *Intracellular Transport*. K. B. Warren (ed.). *Symposia of the International Society for Cell Biology* **5**: 71–103. New York: Academic Press

CRANE, R. K. (1968). A concept of the digestive-absorptive surface of the small intestine. In *Handbook of Physiology. Section 6. Alimentary Canal*. W. Heidel (ed.). **5**: 2535–42

CRANE, R. K. (1969) A perspective of digestive-absorptive function. *American Journal of Clinical Nutrition* **22**, 242–9

CURTIS, A. S. G. (1967). *The Cell Surface: its Molecular Role in Morphogenesis*. London and New York: Logos Press; Academic Press

CURTIS, A. S. G. (1972). Adhesive interactions between organisms. *British Society for Parasitology Symposia* **10**: 1–21

DIKE, S. C. and READ, C. P. (1971). Tegumental phosphohydrolases of *Hymenolepis diminuta. Journal of Parasitology* **57**: 81–87

ERASMUS, D. A. and OHMAN, C. (1965). Electron microscope studies on the gland cells and host-parasite interface of the adhesive organ of *Cyathocotyle bushiensis* Khan, 1962. *Journal of Parasitology* **51**: 761–9

HARLOW, D. R. and BYRAM, J. E. (1971). Isolation and morphology of the mitochondrion of the cestode, *Hymenolepis diminuta. Journal of Parasitology* **57**: 559–65

HOWELLS, R. E. (1965). Electron microscope and histochemical studies on the cuticle and subcuticular tissues of *Moniezia expansa. Parasitology* **55**: 20P–21P

JHA, R. K. and SMYTH, J. D. (1969). *Echinococcus granulosus*: ultrastructure of microtriches. *Experimental Parasitology* **25**: 232–44

JHA, R. K. and SMYTH, J. D. (1971). Ultrastructure of the rostellar tegument of *Echinococcus granulosus* with special reference to biogenesis of mitochondria. *International Journal for Parasitology* **1**: 169–77

JOHNSON, C. F. (1967). Disaccharidase localization in hamster intestine brush borders. *Science* **155**: 1670–72

KWA, B. (1970). Studies on the sparganum of *Spirometra erinacei. M.Sc. Thesis. Australian National University, Canberra*

KWA, B. (1972a). Studies on the sparganum of *Spirometra erinacei*. I. The histology of the sparganum scolex. *International Journal for Parasitology* **2**: (in press)

KWA, B. (1972b). Studies on the sparganum of *Spirometra erinacei*. II. Proteolytic enzymes in the sparganum scolex. *International Journal for Parasitology* **2**: (in press)

KWA, B. (1972c). Studies on the sparganum of *Spirometra erinacei*. III. The fine structure of the tegument in the sparganum scolex. *International Journal for Parasitology* **2**: (in press)

LEE, D. L. (1966). The structure and composition of the helminth cuticle. *Advances in Parasitology* **4**: 187–254

LUMSDEN, R. D. (1966a). Cytological studies on the absorptive surfaces of cestodes. 1. The fine structure of the strobilar integument. *Zeitschrift für Parasitenkunde* **27**: 355–82

LUMSDEN, R. D. (1966b). Cytological studies on the absorptive surfaces of cestodes. II. The synthesis and intracellular transport of protein in the strobilar integument of *Hymenolepis diminuta*. *Zeitschrift für Parasitenkunde* **28**: 1–13

LUMSDEN, R. D., GONZALEZ, G., MILLS, R. R. and VILES, J. M. (1968). Cytological studies on the absorptive surfaces of cestodes. III. Hydrolysis of phosphate esters. *Journal of Parasitology* **54**: 524–35

LUMSDEN, R. D., THREADGOLD, L. T., OAKS, J. A. and ARME, C. (1970). On the permeability of cestodes to colloids: an evaluation of the transmembranosis hypothesis. *Parasitology* **60**: 185–93

MORRIS, G. P. and FINNEGAN, C. V. (1968). Studies of the differentiating plerocercoid cuticle of *Schistocephalus solidus*. I. The histochemical analysis of cuticle development. *Canadian Journal of Zoology* **46**: 115–21

MORRIS, G. P. and FINNEGAN, C. V. (1969). Studies of the differentiating plerocercoid cuticle of *Schistocephalus solidus*. II. The ultrastructure of cuticle development. *Canadian Journal of Zoology* **47**, 957–64

MORSETH, D. (1967). The fine structure of the hydatid cyst and the protoscolex of *Echinococcus granulosus*. *Journal of Parasitology* **53**: 312–25

MUKERJEE, T. M. and WILLIAMS, A. W. (1967). A comparative study of the ultrastructure of the microvilli in the epithelium of small and large intestine of mice. *Journal of Cell Biology* **34**: 447–60

PETERS, W. (1970). Adaptation of the malarial parasite to its environment during the life cycle. *Journal of Parasitology* **57**: No. 4, Section 2, 120–5

RADEMAKER, W. and SOONS, J. B. J. (1957). The distribution of protease activities on liver cell fractions. *Biochimica et Biophysica Acta* **24**: 451–2

READ, C. P. and ROTHMAN, A. H. (1957). The role of carbohydrates in the biology of cestodes. II. The effect of starvation on glycogenesis and glucose consumption in *Hymenolepis*. *Experimental Parasitology* **6**: 1–7

REISSIG, M. (1970). An unusual laminar structure in the integument of *Hymenolepis diminuta*. *Journal of Ultrastructure Research* **31**: 109–15

ROBINSON, G. B. (1963). The distribution of peptidases in sub-cellular fractions from the mucosa of the small intestine of the rat. *Biochemical Journal* **88**: 162–8

ROSARIO, B. (1962). The ultrastructure of the cuticle in the cestodes *H. nana* and *H. diminuta*. In *5th International Congress Electron Microscopy*. S. Bresse (ed.). p.11–12. New York and London: Academic Press

ROTHMAN, A. H. (1963). Electron microscope studies of tapeworms: the surface structures of *Hymenolepis diminuta* (Rudolphi, 1819), Blanchard, 1891. *Transactions of the American Microscopical Society* **82**: 22–30

ROTHMAN, A. H. (1966). Ultrastructural studies of enzyme activity in the cestode cuticle. *Experimental Parasitology* **19**: 332–8

ROTHMAN, A. H. (1967). Colloid transport in the cestode *Hymenolepis diminuta*. *Experimental Parasitology* **21**: 133–6

SIMPSON, C. F. and KLING, J. M. (1968). The mechanism of mitochondrial extrusion from phenylhydrazine-induced reticulocytes in the circulating blood. *Journal of Cell Biology* **36**: 103–9

SJÖSTRAND, F. S. (1967). The structure of cellular membranes. *Protoplasma* **63**: 248–61

SJÖSTRAND, F. S. and BARAJAS, L. (1968). Effects of modifications in conformation of protein molecules on structure of mitochondrial membranes. *Journal of Ultrastructural Research* **25**: 121–55

SMITHERS, S. R., TERRY, R. J. and HOCKLEY, D. J. (1969). Host antigens in schistosomiasis. *Proceedings of the Royal Society,* Series B, **171**: 483–94

SMYTH, J. D. (1964). Observations on the scolex of *Echinococcus granulosus*, with special reference to the occurrence and cytochemistry of secretory cells in the rostellum. *Parasitology* **54**: 515–26

SMYTH, J. D. (1967). Studies on tapeworm physiology. IX. *In vitro* cultivation of *Echinococcus granulosus* from the protoscolex to the strobilate stage. *Parasitology* **57**: 111–33

SMYTH, J. D. (1969). *The Physiology of Cestodes*. London: Oliver & Boyd

SMYTH, J. D. (1971). Development of monozoic forms of *Echinococcus granulosus* during *in vitro* culture. *International Journal for Parasitology* **1**: 121–4

SMYTH, J. D., MILLER, H. J. and HOWKINS, A. B. (1967). Further analysis of the factors controlling strobilization, differentiation, and maturation of *Echinococcus granulosus* in vitro. *Experimental Parasitology* **21**: 31–41

SMYTH, J. D., MORSETH, D. J. and SMYTH, M. M. (1969). Observations on nuclear secretions in the rostellar gland cells of *Echinococcus granulosus* (Cestoda), *The Nucleus* **12**: 47–56

TAYLOR, E. W. and THOMAS, J. N. (1968). Membrane (contact) digestion in the three species of tapeworm *Hymenolepis diminuta, Hymenolepis microstoma* and *Moniezia expansa*. *Parasitology* **58**: 535–46

THREADGOLD, L. T. (1962). An electron microscope study of the tegument and associated structures of *Dipylidium caninum*. *Quarterly Journal of Microscopical Science* **103**: 135–40

THREADGOLD, L. T. (1965). An electron microscope study of the tegument and associated structures of *Proteocephalus pollanicoli*. *Parasitology* **55**: 467–72

THREADGOLD, L. T. and READ, C. P. (1970). *Hymenolepis diminuta*: ultrastructure of a unique membrane specialization in tegument. *Experimental Parasitology* **28**: 246–52

TIMOFEEV, V. A. (1964). Structure of the cuticle of *Schistocephalus pungitii* in the different phases of its development in connection with the specific feeding habits of cestodes. Electron and luminescent microscopy of the cell. (In Russian: English summary). *Zhurnal Obshchei Biologii* **16**: 50–60. Nauka: Moscow and Lennigrad. 1965

UGOLEV, A. M. (1965). Membrane (contact) digestion. *Physiological Reviews* **45**: 555–95

UGOLEV, A. M. (1968). *Physiology and pathology of membrane digestion*. New York: Plenum Press

WAGNER, R. P. (1969). Genetics and phenogenetics of mitochondria. *Science* **163**: 1026–31

WEINBACH, E. C. and BRAND, TH. VON. (1970). The biochemistry of cestode mitochondria. I. Aerobic metabolism of mitochondria from *Taenia taeniaeformis*. *International Journal of Biochemistry* **1**: 39–56

YAMANE, Y. (1968). On the fine structure of *Diphyllobothrium erinacei* with special reference to the tegument. *Yonago Acta medica* **12**: 169–81

YAMANE, Y. (1969). An electron-microscope study of the tegument of *Diplogonoporus grandis*. *Yonago Acta Medica* **13**: 25–9

THE HOST-PARASITE INTERFACE OF PARASITIC PROTOZOA.

SOME PROBLEMS POSED BY ULTRASTRUCTURAL STUDIES

KEITH VICKERMAN

Department of Zoology, University of Glasgow

The protozoan parasite surface provides the portals for traffic of substances into and out of the parasite, the sites of combat with host defences, and in many cases the mechanisms for parasite movement and attachment to the host, even for penetration into a host cell preparatory to leading an intracellular existence. My theme will centre about these properties of the parasite's surface solely in relation to the host environment, but I shall take as a starting point the structure of the host parasite interface as seen with the electron microscope (EM) and go on to deal with the interpretation of ultrastructural findings in terms of parasite function.

First, let me say a few words about the endless variety of form which we encounter when we look at the outermost layers of protozoan parasites. Like all other eukaryote cells, protozoa are bounded by a three-ply unit membrane (8–10 nm thick) as seen in section at the EM level. This membrane is usually reinforced by cortical structures—other membranes, filaments or microtubules —which support the shape of the body and constitute the so-called pellicle. Extracellular stages of sporozoans [sporozoites, merozoites, trophozoites of gregarines (Plate 1*)] for example, have other membranes (reviewed by Vivier *et al.*, 1970) lying below the plasmalemma: these may be discontinuous in some cases (e.g. merozoites of *Plasmodium fallax*, Aikawa, 1967). Ciliates (Plate 2) and dinoflagellates have membranous alveoli in the same position. Trypanosomes, trypanoplasms and those stages in the sporozoan life-cycle which move by flexion of the body have a layer of pellicular microtubules (Plates 1A, B; 2A; 3A, B; 4A; 5B). In the sporozoa such microtubules are believed to play a part in the flexing movements (reviewed by Garnham, 1966).

The whole surface layer may be elaborately convoluted in non-phagotrophs as in eugregarines (Plate 1C) and the more bizarre zooflagellate symbionts of insects. Scanning electron microscopy (Vavra and Small, 1969) has demonstrated convincingly that the interaction of surface folds can provide a mechanism for jet propulsion in eugregarines. The elaborate surface folds of the

---

* Plates appear between pages 72 and 73.

flagellates *Streblomastix strix* (Grimstone, 1961), *Mixotricha paradoxa* (Cleveland and Grimstone, 1964) and *Lophomonas striata* (Beams *et al.*, 1960) serve as increased surface area for the attachment of bacterial symbionts. The undulating membrane of trypanosomes, trichomonads and the archigregarine *Ditrypanocystis* (Macgregor and Thomasson, 1965) can be regarded as increased surface area to provide additional locomotory thrust in the viscous media inhabited by these parasites.

In this short review I shall discuss some problems posed by the host-parasite interface of protozoa which are (a) freely moving in an extracellular milieu, (b) attached to a host substratum, and (c) invading and living inside a host cell.

EXTRACELLULAR PROTOZOA

Outside the surface unit membrane of many animal cells electron microscopy has disclosed a layer of material forming a surface coat. In some cases—for example the microvillar surface of gut epithelial cells or the plasmalemma of large free-living amoebae, this coat is rich in carbohydrate and is termed the glycocalyx (reviewed by Bennett, 1969). We have considerable evidence that this coat has a role in assisting the adhesion of the cell either to its substratum or to other cells, in the binding of particles for pinocytosis or phagocytosis, and in determining the antigenic character of the cell. Sceptics have pointed out that in many cases this coat may be a fixation artifact—the result of precipitation of foreign material on the cell's surface or of the leaching out of cytoplasmic colloids. This may well be so, but in other cases well documented autoradiographic studies have demonstrated the production of cell coat material in the cell's synthetic machinery, packaging of coated membrane in the Golgi apparatus and interpolation of these packages into the surface layer.

Surface coats have been described for many protozoan parasites. The coat may cover the entire surface or be restricted to particular regions. *Trichomonas vaginalis* has a fuzzy coat only at focal sites on the surface where binding of colloidal particles for pinocytosis occurs, elsewhere the plasmalemma is smooth (Nielsen, 1970).

Entamoeba histolytica from the colon has a diffuse surface coat, but this is lacking from amoebae grown in axenic culture (El-Hashami and Pittman, 1970; Griffin and Juniper, 1971)—a difference which may be significant in considerations of pathogenesis. Damage to host cells involves physical contact (Jarumilinta and Kradolfer, 1964) but too little attention has been paid to the way in which the amoeba attacks the gut wall. Prathap and Gilman (1970) have shown that the primary site of attack is the epithelium rather than the crypts and Eaton and co-workers (1970) have suggested that the rupture of a surface lysosome is responsible for lysing the epithelial cell and so starting the invasion.

N.B. All illustrations are electron micrographs of sections. Material was fixed in glutaraldehyde and post-osmicated unless otherwise stated.

T.S.—transverse section   L.S.—longitudinal section

d	desmosome		N	nucleus
f	flagellum		om	outer membrane
fv	food vacuole		omh	outer membrane (host)
im	inner membrane		P	parasite
HC	host cell		pmt	pellicular microtubule
HCN	host cell nucleus		rib	ribosomes
m	mitochondrion		sc	surface coat
mh	mitochondrion (host)		sch	surface coat (host)
mvh	microvilli (host)		sm	surface membrane
			twh	terminal web of host epithelial cell

PLATE 1

Surface features of gregarines from polychaetes.

A. T.S. of *Selenidium* sp. from *Sabellaria lapidaria*. Note 2 inner membranes (im1, im2) beneath surface membrane, and cortical microtubules. × 72,000. (Courtesy of Dr Gundula Dorey.)

B. L.S. anterior region of *Selenidium hollandei* from *Sabellaria alveolata*. Note gap between parasite mucron (mu) and host cell (arrowed), absence of attachment plaques, ingestion of material through conoid (c) into food vacuole, pellicular microtubules in glancing section, and profiles of rhoptries (r). × 30,000. (After Schrevel, 1968.)

C. L.S. of mucron (mu) of *Lecudina pellucida* attached to gut cell of *Perinereis cultrifera*. Note pyknotic cytoplasm of this cell in comparison with that of cells on either side. Within the mucron tracts of fibrils (6–7 nm) converge on the dense attachment plaque, details of which are shown in relation to the opposed membranes at the arrowheads. The fine fibrils may be associated with pinocytosis which takes place around the periphery of the mucron. Such fibrils are absent from the epimerite of cephaline gregarines. Pellicular ridges (pr) are characteristic of gregarines. × 24,000. (After Schrevel and Vivier, 1966.)

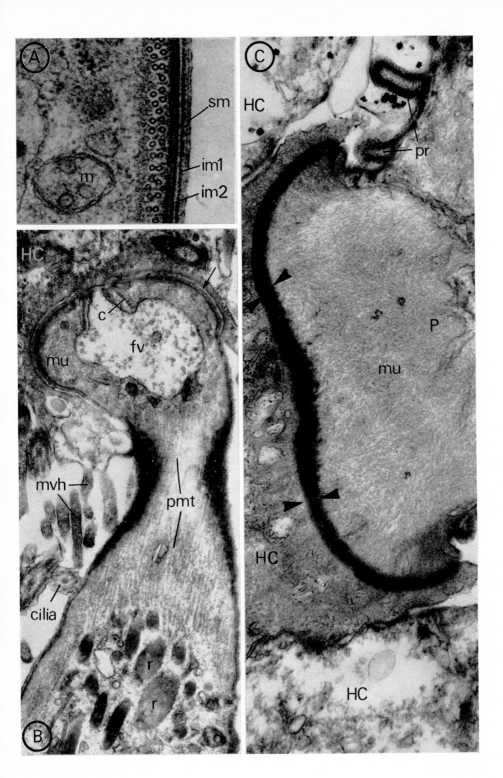

PLATE 2

Attachment devices of protozoa on gills of freshwater fishes. (Courtesy of Dr J. Lom.)

A.   *Cryptobia branchialis* (kinetoplastid flagellate). T.S. of trailing flagellum to show attachment to host cell (between arrows). The latter's surface coat is obliterated along the line of flagellar attachment, but there is no obvious reorganisation of fibrils in the host cell's terminal web to indicate a desmosome attachment. Arrowheads delimit the region of attachment of the parasite's flagellum to its body (macular desmosomes studding this region are not shown in this section). × 62,000.

B.   *Oodinium cyprinodontum* (dinoflagellate). Note desmosome-like attachments of parasite to host cell with extensive realignment of host cell's fibrils and microtubules to converge on these attachments.   × 44,000.

C.   *Trichophrya piscium*. This suctorian has surface alveoli (alv) between which lie pores (white arrows) secreting a cementing substance (cem). This cement bonds with the host cell's surface coat to form a firm attachment. Fine helices (black arrows) are present in the cement. There is no obvious reorganisation of the adjacent host cell. × 55,000.

D.   *T. piscium*. Enlarged view of helical structure which traverses the cementing layer between surface membranes of parasite and host. × 153,000.

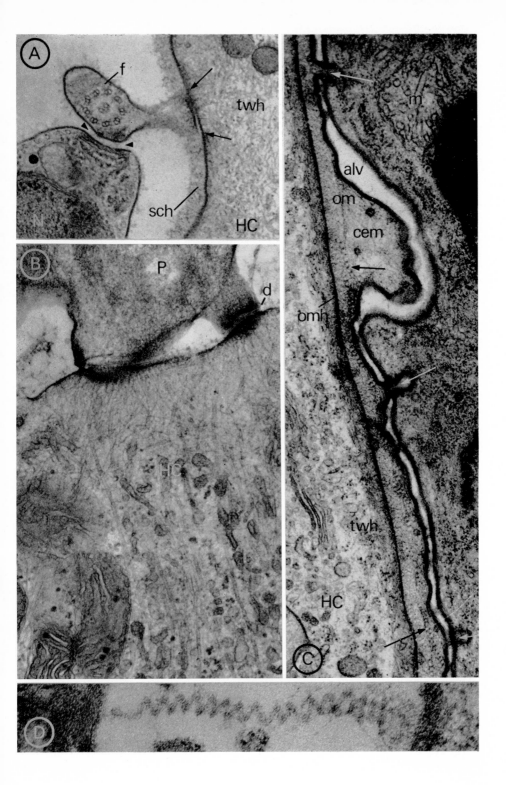

PLATE 3

A. *Trypanosoma brucei*. Part of T.S. of bloodstream form treated with homologous ferritin-conjugated antiserum. The ferritin conjugate particles (arrowed) adhere to the thick surface coat which overlies the surface membrane of body, flagellum and plasmanemes (pl). Formalin fixation. × 150,000.

B. *T. brucei*. Part of T.S. of trypanosomes in tsetse midgut cardia. The surface coat is missing. The gap between flagellum and body is approximately the same as that between the bodies of adjacent flagellates. Arrowheads indicate position of desmosome (not in plane of this section) linking body and flagellum. × 72,000.

C. *Histomonas meleagridis*. Section through large pseudopodium (ps) of invasive stage in the turkey. The parasite is phagocytosing (at white arrows) portions of adjacent host cells including an heterophil with prominent granules (hg). Note the gap between the thick parasite surface membrane and the thin host cell membrane (at arrowheads). × 20,000. (After Lee *et al.*, 1969.)

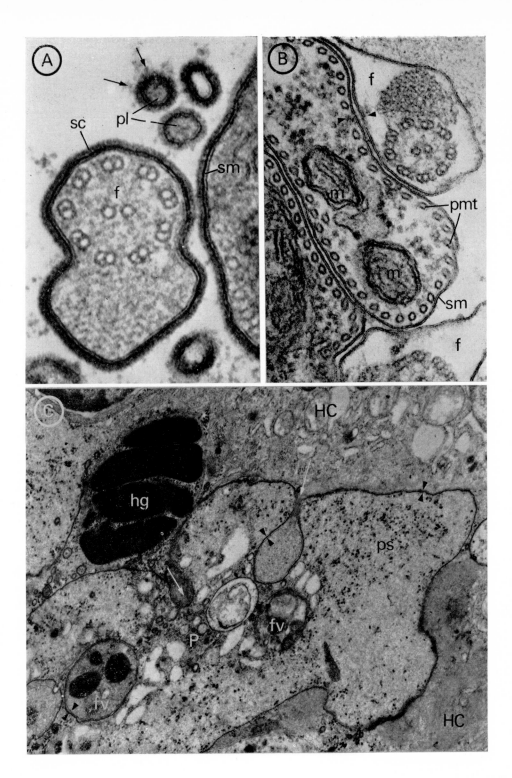

PLATE 4

L.S. of *Crithidia fasciculata* attached to the cuticular lining (cut) of the rectum of *Anopheles gambiae*. Note desmosomal junctions between flagellum and wall of flagellar pocket (fp), also hemidesmosomal junction where the truncated flagellum is attached (between arrowheads) to the cuticle. A mass of filaments (arrowed) ending at a dense plaque is common to both types of junction, so is a filamentous intermembraneous matrix. $\times$ 60,000. (Courtesy of Mr B. E. Brooker.)

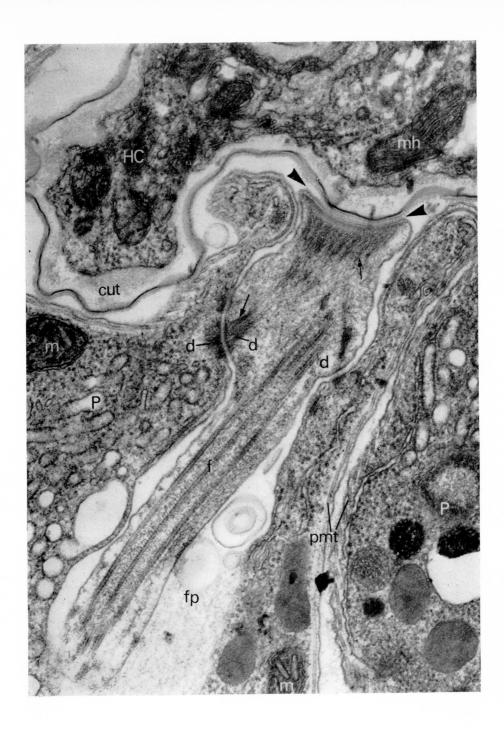

PLATE 5

Sections to show penetration of host cells by merozoites.

A. *Cryptosporidium wrairii*. Merozoite entering microvillar border of epithelial cell in guinea pig ileum. The host cell surface is drawn up into a cup (chcs) around the invading parasite. A thin non-membraneous abscission layer (abl) segregates the site of penetration. × 25,000. (After Vetterling *et al.*, 1971.)

B. *Plasmodium gallinaceum*. Two merozoites penetrating a chicken red cell. The upper parasite is shown invaginating the host cell in the direction of the arrow (the microtubules converge on the apical intratorium which is just out of the plane of section). Note the fuzzy coat which may be shed by the parasite on entry. The lower parasite has completed entry and has begun to dedifferentiate (breakdown of apical complex and internal compound membrane (icm)). × 60,000. (After Ladda *et al.*, 1969.)

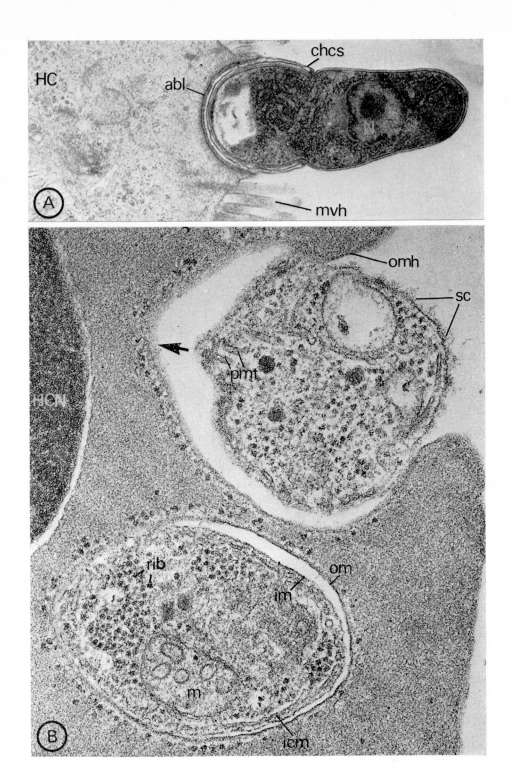

It is tempting to speculate that the various factors cited as encouraging the transition from commensal to pathogenic phase may exert their influences by abrading the host epithelial cell's protective glycocalyx, so enabling the amoeba to gain purchase on the cell's plasma membrane and start its career of destruction.

But the most controversial of parasite surface coats among protozoa are to be found on trypanosomes, and it is with the relation of these coats to the ability of these flagellates to evade the host's immune assault that I shall be largely concerned in this section.

Electron micrographs of pathogenic (Salivarian) trypanosomes taken from the mammalian bloodstream show that a smooth and compact coat (Plate 3 A) envelops the entire surface membrane of body and flagellum (reviewed by Vickerman, 1971). This coat is lost when trypanosomes commence cyclical development in the tsetse fly (Plate 3 B) or in culture, but in *Trypanosoma brucei* (other species have not yet been investigated) the coat is regained by the metacyclic trypanosomes. This cyclical coating and uncoating can be correlated with cyclical changes in the net surface charge on the trypanosome (Hollingshead *et al.*, 1963; Lanham, 1968).

The coat has been interpreted as:

(a)   an endogenous secretion related to the trypanosome's defences against the host's immune response, in particular to the phenomenon of antigenic variation shown by these parasites and extensively investigated for *T. brucei* (Vickerman, 1969b),

(b)   an exogenous accretion of host serum protein involved in disguising the trypanosome as host (as described for schistosomes by Smithers and colleagues, see Clegg p. 29) or composed of adsorbed host antibody (Desowitz, 1970),

(c)   a combination of (a) and (b) (Vickerman, 1971).

I shall deal with each of these interpretations in turn.

Trypanosome populations isolated from successive relapses in the bloodstream differ in their surface antigenic character as shown by agglutination and neutralization reactions (reviewed by Gray, 1967). This succession of antigenic types (A → B → C → D etc.) in chronic infections, even those initiated as clone strains, has been explained as resulting either from selection of genetic mutants (Watkins, 1964) or from induction of antigenic change in individual trypanosomes (Gray, 1967). But each cyclically transmitted strain follows a semi-predictable sequential pattern of variation in different hosts, starting from a 'basic antigen' to which the trypanosomes revert each time they are passed through the tsetse fly (Gray, 1965). This important observation militates against the mutation/selection theory.

Weitz (1960a, b) noted that bloodstream *T. brucei*, when very well washed, are not agglutinated by variant-homologous antiserum, and that trypanosomes apparently release their variant antigens into the blood as soluble 'exoantigen'. Such washing also removes the trypanosome's surface coat, and variant-specific antibody when conjugated to ferritin (Plate 3 A) will not adhere to de-coated trypanosomes (Vickerman and Luckins, 1969), suggesting that the variant antigen is located in the coat. Culture trypanosomes derived from different bloodstream variants show a common surface agglutinogen (Seed, 1964). This loss of antigenic identity in culture (and by analogy in the fly mid-gut) may be related to loss of coat. In the salivary glands of *Glossina* the meta-cyclic trypanosomes reacquire the coat and such trypanosomes possess the 'basic antigen' of the first bloodstream population, as shown by neutralization reactions (Cunningham, 1966).

Early studies on the nature of the variant antigens suggested that they were unconjugated proteins with a sedimentation coefficient of $4S$ (Williamson and Brown, 1964). The surface coat of fixed trypanosomes can be digested away with proteolytic enzymes (Vickerman, 1969b; Wright and Hales, 1970). More recent studies using isoelectric focussing in polyacrylamide gels (Allsopp *et al.*, 1971) indicate that the $4S$ antigen has at least 3 components with a common antigenic determinant. Each component is a protein-carbohydrate complex with an isoelectric point (pI) in the pH range 6·0–5·5. The $4S$ antigen is not continuously shed by trypanosomes into the bloodstream as exoantigen, for it cannot be detected in recently collected plasma but appears in increasing quan-tities when infected blood is allowed to stand on the bench. A drop in pH appears to signal discharge of the exoantigen as trypanosomes buffered at pH 8·0 do not release it. The mechanism of release has been revealed by EM studies on negatively-stained trypanosomes (Wright *et al.*, 1970): the trypanosomes shed their surface membrane *plus* coat as long streamers ('Filopodia' *sensu* Wright *et al.*, 1970; plasmanemes *sensu* Vickerman and Luckins, 1969). These streamers (Plate 3A) are immunogenic (Herbert and Macadam, 1971).

The demonstration of a carbohydrate component in variant antigen and the location of this antigen in the surface coat is supported by cytochemical studies of Wright and Hales (1970). Using periodic acid-silver staining of material sectioned for electron microscopy, they located the carbohydrate layer at the junction of the surface coat and the outer leaflet of the surface membrane, and comparison with non-stained images suggested that the layer formed the innermost part of the coat. Trypsin digestion, while leaving the surface mem-brane intact, removed the carbohydrate layer along with the surface coat. Carbohydrates figure prominently in the composition of the surface coat (glycocalyx) of many other cells including that of free-living flagellates (for references see Vickerman, 1969a). Detailed autoradiographic studies have shown that the coated membrane is synthesized in the cell's Golgi apparatus. I have argued in favour of a similar origin for the surface coat of trypanosomes (Vickerman, 1969a), as all the machinery of secretion is present close to the flagellar pocket of bloodstream trypanosomes and the entire apparatus is much reduced in culture forms.

But what of the alternative origin of the trypanosome surface coat as adsorbed host serum protein? As a surface coat is present in trypanosomes from non-immune or immunosuppressed hosts (Vickerman, unpublished observations) and adsorbed antibody is not usually visible on cells following conventional preparation for electron microscopy, the likelihood of the coat being antibody seems remote. But Strauss (1971) has published some thought-provoking electron micrographs of culture forms of *Leishmania tarentolae* grown in the presence of homologous antibody (diluted 1 : 100–1 : 1200). A globular surface coat covered the flagellates which formed pseudosyncytial masses with apposed surface coats separated by a dense band. Both coat and syncytia were absent from flagellates grown in normal serum. The distance between two agglutinated cells was said to be greater than the size of either an IgG or an IgM molecule, so if the electron dense coat does represent antibody the globulins may be bound to an otherwise undetected surface coat present in culture flagellates.

From their work on the role of sheep serum supplement in adapting *Trypanosoma vivax* to rodents, Desowitz and Watson (1953) suggested that trypanosomes might adapt to new host species by binding specific plasma proteins to their surface. More recently, Desowitz (1970) discussed the possibility that such proteins might compose the trypanosome's surface coat.

Ketteridge (1971) found that washed *T. vivax* from mice could be agglutinated by rabbit anti-mouse-blood serum at high titres (1 : 1000) and agglutination could be prevented by absorption of the antiserum with mouse serum proteins, more specifically the α- and β-globulin fractions. Agglutination was followed by lysis in the absence of added complement. The infectivity of mouse-derived *T, vivax* to mice was impaired by pre-incubation of the trypanosomes in the anti-mouse serum at dilutions beyond the lytic titre. These observations suggest that host-serum proteins are bound to the surface of the trypanosome and that the trypanosome provides its own complement factors for lytic reactions with anti-host antibodies. Mouse-derived *T. brucei* could also be agglutinated by anti-mouse serum, but lysis did not occur, even on the addition of complement, so the association between host serum components and *T. brucei* is not so intimate as to involve the parasite's surface in the effects of complement interactions.

From studies on the elution characteristics of trypanosomes passed through columns of ion exchange resins, Lanham (1968) concluded that *T. vivax* carried a greater density of negative charges than *T. brucei*. Positively-charged colloidal iron hydroxide particles bind to the surface of red blood cells but treating the cells with neuraminidase prevents such binding, as the anionic sialic acid is released from glycoproteins in the cell membrane with concomitant loss of negative charge (reviewed by Cook, 1968). Bloodstream *T. vivax* also shows neuraminidase-sensitive binding of colloidal iron but bloodstream *T. brucei* does not bind the colloid at all (Ketteridge, 1971). Precipitin reactions between anti-*T. vivax* serum and starch-gel electrophoretograms of normal mouse serum indicate that the antigen common to parasite and host lies in the faster-moving α_2-glycoprotein region of electrophoretic mobility.

Acid glycoproteins are believed to be capable of exerting 'colloid protective capacity' and of preventing the complexing of antigens with circulating antibody (Apffel and Peters, 1970). The serum glycoproteins adsorbed by rodent-adapted *T. vivax,* may therefore serve to protect it against host antibodies. In sheep this trypanosome, like *T. brucei,* evades host antibodies by undergoing antigenic variation (Clarkson and Awan, 1968), but the rodent adapted strain is too virulent to exhibit such variation in untreated infections.

Desowitz and Watson (1953) believed that *T. vivax* in sheep might bind serum components, and that these components might protect the trypanosomes against homologous antibody *in vivo. In vitro,* they found that trypanosomes were lysed by homologous sheep antiserum only if they were separated from the blood and washed before the reaction. Sheep serum components may not, therefore, bind so tightly to the trypanosome surface as mouse serum components in the rodent adapted strain. Although not so well worked out, then, it appears that *T. vivax,* like *Schistosoma mansoni* (see p. 29), can bind serum proteins to its surface, and such proteins may reduce the opsonising activity of host antibodies.

Perhaps the most acceptable interpretation of the surface coat of blood-stream salivarian trypanosomes, then, is the third one—i.e. that it is compounded of endogenous secretion (which includes variant antigens) and an exogenous accretion of host serum proteins. The relationship between these two may be somewhat different in *T. brucei* and *T. vivax* despite the similarity in form and dimension of the coats in these two species.

Bloodstream *T. lewisi* presents an entirely different coat in electron micrographs (Vickerman, 1969b)—tattered and diffuse rather than dense and compact, but this too disappears in culture. The rat trypanosome is even more negatively charged than *T. vivax* and like the latter it binds the serum proteins of its host with avidity (D'Alesandro, 1966). The 'exoantigens' of *T. lewisi* are complexed with host serum proteins but whether the exoantigens are temporarily secreted onto the flagellate's surface before they are released is not known (D'Alesandro, 1970).

Rat serum β-globulins promote the proliferation of *T. lewisi* in the mouse (Lincicome, 1958; Greenblatt *et al.,* 1969) which is normally refractive to this parasite. This growth-stimulating activity is the opposite of what happens in the rat 7 days after infection, when cessation of multiplication is thought to be brought about by the antibody 'ablastin'. Both active principles have comparable electrophoretic mobilities and molecular sizes, and neither can be absorbed out by excessive numbers of trypanosomes. In the rat competition for surface receptor sites may determine whether the parasite ceases to multiply or not.

Patton (1970—personal communication) has recently studied the mode of action of ablastin and suggests that it blocks the ATP-activated Na^+/K^+ pump. The resulting dearth of potassium in the cell could account for the cut-back in both protein synthesis (by limiting translation) and glycolysis that has been observed at this stage in the life-cycle. In this one example we can see that further work on the effect of host antibody on specific transport mechanisms might add considerably to our understanding of how parasite differentiation may be controlled by the host.

ATTACHED PROTOZOA

Attachment to host cells plays an important part in the life-cycles of many parasitic protozoa, especially those that undergo growth or multiplication in host cavities where the surrounding medium is subject to movement. Differentiation (i.e. progression to the next stage in the life-cycle) is often associated with detachment. The trypanosomatids and gregarines afford good examples.

Specializations of the cell periphery related to adhesion and attachment in metazoa have been subjected to considerable experimental study as outlined by Curtis (p. 1) in this symposium. The attachment devices of protozoan parasites have received no such treatment. As yet the ultrastructural aspects of these attachments are poorly known, but even with only fragmentary knowledge we can begin to make comparisons, and several questions spring to mind. Do attached protozoa show a spectrum of cell junctional form similar to that of higher organisms (see Kelly and Luft, 1966)? Do taxonomically unrelated protozoa attached to the same tissues show similar junctional complexes? How do these complexes arise and what makes them disappear? How far is host specificity related to the ability to develop junctional complexes? Are these complexes related to parasite size or to host cell damage in pathogenic protozoa?

Insofar as I am aware, most protozoan-host cell contacts reported to date have been in the 'secondary minimum' of Curtis (1967). Tight junctions do not occur but a wide range of attachments of the 'intermediate junction' and desmosome (*macula adherens*) types arise around the 10–20 nm gap. A similar gap is found between motile parasites in tissues and their host cells e.g. in the invasive stage of *Histomonas meleagridis* (Plate 3C, Lee *et al.*, 1969).

Desmosomes (reviewed by Campbell and Campbell, 1971) are characterized by dense plaques on either side of the apposed membranes, with bundles of fine fibrils (tonofilaments) converging on the plaques. Between like cells they are usually symmetrical but between cells at different stages of differentiation they may be asymmetrical. Protozoa and their host cells are more than developmentally different—they are genetically poles apart, so it is not surprising that in many of their attachment devices asymmetry is the rule.

Despite the fact that a successful means of preventing attachment in the vector could lead to the elimination of cyclically-transmitted trypanosomes, little attention has been paid to the mechanism of trypanosome attachment. We have most information on the insect trypanosomatid, *Crithidia fasciculata* (Brooker, 1970, 1971a, 1971b). Among the trypanosomatids we find a series of desmosome-like plaques between the flagellum and body in trypanosomes and between flagellum and flagellar pocket wall in crithidias (Plate 4). In culture the flagellates of *C. fasciculata* adhere to one another by their flagella and in electron micrographs desmosomes may be found between adjacent flagella. Alternatively, rosettes of flagellates may be formed around membranous flagellate debris and the flagella develop hemidesmosomes at the site of attachment (Brooker, 1970). Similar hemidesmosomes have been found where flagellates have become attached to Millipore filters *in vitro* (Brooker, 1971a) and in the mosquito host where the so-called haptomonad forms attach by their shortened flagella to the chitinous cuticle lining the fore- or hind-gut. Detachment of flagellates from their substrate, or from one another, and assumption of the swimming (necto-monad) form can be induced by distilled water. Such treatment results in the area of each hemidesmosome contact becoming reduced by endocytosis of membrane. At the same time there occurs loss of the internal filaments, so that when the flagellum starts to beat, separation is easily accomplished. A similar story has been told for the dissolution of desmosomes in disaggregated chick cells.

It is interesting that although flagella produce hemidesmosomes in contact with a variety of substrates (cellulose, chitin, membranous debris), desmosomes are found only between adjacent flagella and not between body and flagellum of adjacent flagellates. In most cases trypanosomatid attachment is to a chitinous surface and hemidesmosome-like structures are to be expected (c.f. Molyneux, 1969, for *T. lewisi* in flea hindgut).

Cases in which trypanosomes attach directly to living host cell surfaces have been poorly studied. In *T. brucei* attached to tsetse salivary gland epithelium (Vickerman, unpublished observations) the flagella appear simply to intertwine with epithelial microvilli.

The development of attachment structures in gregarines has been beautifully documented at the EM level by Desportes (1969), Schrevel (1968) and Schrevel and Vivier (1966) and reviewed by Ormieres and Daumal (1970). In these accounts we see varying degrees of increased density of the finely fibrous cytoplasm alongside apposed membranes (Plate 1 C) of mucron and host cell, but not of epimerite and host cell. Early descriptions of the attachment structures of gregarines suggested, in many cases, elaborate rootlet systems penetrating the host cell. Thus Hesse (1909) portrayed fibrils traversing the junction of the *Nematocystis* trophozoite and the host cell promontory to which it is anchored. Macmillan (1969), using the polarizing microscope on the parasite, has confirmed the existence of the fibrillar organization but we still await EM studies to show that these in fact represent an elaborate desmosome. *Grebnickiella* has long rhizoids apparently penetrating host epithelial cells but Tuzet and Galangau's (1968) electron micrographs show that the branched radicels end in blunt processes resembling, on a small scale, the mucron of other gregarines. These processes abut on the invaginated host cell membrane but do not penetrate it—though trans-junctional effects of the parasite can be seen in the remarkable stacking of the host cell granular reticulum perpendicular to the radicels. Dissolution of the contents of the host cell beyond the junctional complex is seen in gregarine attachment, and in archigregarine trophozoites (Plate 1 B) Schrevel (1968) has found ingestion through the persistent conoid of the trophozoite. In acephaline gregarines there is some indication of pinocytosis taking place around the edge of the mucron and the fine fibrils of this structure (Plate 1 C) may have some function in both mucron contractility and pinocytosis (c.f. Allison *et al.*, 1971). In the cephaline gregarines fine fibrils are absent and no such ingestion occurs (Schrevel and Vivier, 1966) but the possibility of contact digestion occurring at the junction should be investigated. Gregarine nutrition is now more of a mystery than gregarine movement!

Ectoparasitic protozoa are among the most pathogenic parasites of fishes, especially those attaching to the gills. The taxonomic variety of these parasites, living in the same situation, makes comparative study of the specificity and structure of their attachment devices of particular interest and such a study is being undertaken by Lom and colleagues (Lom and Corliss, 1970; Lom and Lawler, 1971—personal communication; Lom, 1971). The small flagellate *Cryptobia branchialis* is attached to the gill epithelial cell by its trailing flagellum. Despite the thick surface coat of the host cell, the membranes of parasite and host are separated by the usual gap and no desmosome-like structures are visible, though, as in the related trypanosomatids, a series of macular desmosomes connect this flagellum to the parasite's body. Yet the flagellate remains attached in the face of strong water currents over the gill filaments. The larger dinoflagellate *Oodinium cyprinodontum*, causes realignment of the fine fibrils of the host cell's terminal web to form prominent asymmetrical desmosomes at points of contact with the parasite. In the still larger peritrich ciliates *Apiosoma* and *Epistylis* orientated fibrils of both parasite stalk and host cell converge at small maculae on the apposed membranes—giving the nearest approach to a symmetrical desmosome as found in higher organisms. The suctorian *Trichophrya piscium* secretes a cement which appears to bond with the epithelial cell's surface coat to form a firm attachment. Helical filaments (Plate 2 D) traverse this cementing material but as in all other cell junctions there is no evidence that the filaments cross membranes and the host cell alongside shows no internal reorganization.

None of the above gill dwellers appears to cause dissolution of the host cell as do gregarine trophozoites, despite the variety of junctional complexes encountered. Other fish ectoparasites show haptor-like devices somewhat reminiscent of those of helminths. Thus *Trichodinella* (Peritricha) has a smooth adhesive disc, the sharp borders of which constrict portions of the epithelium causing damage to the gills. A more famous 'adhesive disc' is that of the common gut flagellate *Giardia*—but alas it appears to be wrongly named! The EM studies of Friend (1966) suggest that it is a rigid structure, supportive rather than contractile. It is the peripheral flange of the ventral surface that appears to grasp the host's intestinal villus and the paramyosin-like substructure of the flange's interior lends credibility to the idea that this is a contractile structure. How *Giardia* feeds is not obvious, but Solovjev and Platova (1969) have pointed out that the distribution of the parasite along the intestine and along individual villi may be related to the pattern of distribution of enzymes involved in contact digestion in the brush border. Raised on its pontoon above the sites of digestive activity, the flagellate's ventral surface can readily absorb the products of digestion which are wafted over it by the action of the ventral flagella.

INTRACELLULAR PROTOZOA

The relationship of host and parasite membranes is of particular interest in those protozoa that penetrate a host cell and develop therein. EM studies of these protozoa have presented us with some fascinating problems in membrane biology—as to how membranes form, grow, fuse and change their composition. I shall be concerned mostly with how the method of entry adopted by the parasite may influence the intimacy of its subsequent relationship with the host cell.

In the microsporidian life-cycle the sporoplasm is injected into the host cell. Spore dehiscence is accompanied by extrusion of the polar filament which penetrates the host cell membrane and swelling of the polaroplast within the spore results in internal pressure increases which propel the sporoplasm out through the filament. Inside the host cell the trophozoite is separated from the host cell cytoplasm by one membrane. Vavra (1965) believed this to be of parasite origin but Sprague and Vernick (1968) maintained that it is continuous with host cytoplasmic membranes and therefore must be of host origin. These authors believed that the microsporidian germ is subcellular, that the dehiscing spore injects only its genetic material around which a new trophozoite is organized. This view of microsporidians as 'eukaryotic phages' is fascinating but improbable!

Broadly speaking, all other parasitic protozoa enter the host cell by one of two methods—by boring through the cell membrane and allowing the host to repair the breach, or by becoming engulfed in a vacuole by the host cell. Lee

(1969) believed *Histomonas* to enter the oocyte of *Heterakis* by the first method as, once inside, the parasite is surrounded by just one membrane—its own—around which cytolysis occurs. Those protozoa that develop in macrophages obviously rely on the second method. *Leishmania* promastigotes are ingested into a membrane-lined vacuole which shrinks during parasite transformation (Miller and Twohy, 1967) until eventually the amastigotes are surrounded by two membranes (Rudzinska *et al.*, 1964), one of host origin. But a study of how intracellular sporozoan parasites gain access to the host cell's cytoplasm shows that these examples are oversimplifying the problem.

Despite the superficial similarity in structure of the infective stages of sporozoa (reviewed by Porchet-Hennere and Vivier, 1971), different mechanisms of cell penetration are said to operate for those sporozoites and merozoites which possess a conoid (coccidians, including *Toxoplasma* and its like) and those that do not (haemosporidians and piroplasms).

Hammond and his collaborators (1971) have described the penetration of sporozoites of various coccidians into host cells. The sporozoites enter with no marked deceleration of their gliding movement and entry takes only a few seconds. As it enters the cell the sporozoite contracts as though passing through a very small opening in the cell membrane. The apical conoid may play a part in penetration as it appears to be extrudable and retractable (McLaren and Paget, 1968; Ryley, 1969; Roberts and Hammond, 1970). The rhoptries (or paired organelles) which open through the conoid have been envisaged as secreting lytic enzymes which might produce a breach in the host membrane (McLaren and Paget, 1968; Jadin and Creemers, 1968), but the evidence for a lysosomal function of these sac-like structures in any sporozoan so far rests only on the positive acid phosphatase activity of rhoptries in the trophozoite of the gregarine *Selenidium* (Schrevel, 1968).

Ladda *et al.* (1969) considered the orientation of *Plasmodium* merozoites towards the red cell to be passive. Although entry was again apex first, a protrudable conoid is absent from malaria parasites, and Scholtyseck *et al.* (1970) have proposed the term 'intratorium' for the complex of reinforcing polar rings sometimes referred to as the conoid in these organisms. During penetration the red cell membrane is believed to remain intact and to become invaginated about the invading parasite (Plate 5). A decrease in size and density of the rhoptries during penetration was thought by these workers to reflect expenditure of their contents, but they may produce a surface active agent which could facilitate expansion of the erythrocyte membrane.

There are, however, exceptions to the above dichotomy. The conoid-bearing merozoite of the quasi-coccidian *Cryptosporidium* (Vetterling *et al.*, 1971) and the conoid-bearing sporozoites of the intracellular gregarine *Lankesteria* (Sheffield *et al.*, 1971) most certainly invade the brush border of intestinal epithelial cells by invaginating the host cell membrane (Plate 5 A).

The intracellular trophozoites of coccidians, gregarines and malaria parasites are surrounded by two membranes, an inner one of obvious parasite origin
and an outer membrane, lining what is often termed the parasitophorous
vacuole. But where this second membrane comes from has been a matter of
dispute (e.g. see Aikawa *et al.*, 1969, and Rudzinska, 1969). It may be derived
directly from the host cell membrane, it may be new membrane formed around
the parasite after penetration, or it may in some cases be a second membrane
produced by the parasite itself on entry.

The first possibility is supported by Ladda *et al.* (1969) for malaria parasites
(Plate 5), Vetterling *et al.* (1971) for *Cryptosporidium*, and Sheffield *et al.* (1971)
for *Lankesteria*, from a study of electron micrographs of penetrating merozoites
or sporozoites, and from the observation that the second membrane is missing
from malaria parasites cultured *in vitro* (Aikawa *et al.*, 1969). This makes sense
if the parasite actually invaginates the cell membrane. For the conoid-bearing
coccidians (other than *Cryptosporidium*), however, the presence of an incomplete
outer membrane around recently penetrated sporozoites (Scholtyseck, 1969;
Roberts *et al.*, 1970) provides evidence for the second possibility, not to mention
the prolific shedding of this membrane as streamers into the space between outer
and inner membranes of the parasite (Hammond *et al.*, 1967).

If the outer membrane of the mammalian malaria parasite is of host cell
origin then both entry and subsequent growth of the parasite would necessitate
synthesis of new membrane by the anucleate red cell. The possibility that the
outer membrane is of parasite origin seems remote, but it is supported in the
case of certain malaria parasites by the observed continuity of this membrane
with membrane-bound clefts or vacuoles in the host cell cytoplasm (Rudzinska
and Trager, 1968). These correspond to the Maurer's clefts and Schuffner's dots
of light microscopists and studies with fluorescein-conjugated anti-parasite sera
show that the clefts fluoresce along with the bodies of the parasites, indicating
that they contain parasite antigens (Tobie and Coatney, 1961; Voller, 1965).

In view of the controversy over the origin of the outer membrane in coccidians and haemosporidians, it is all the more intriguing to find that it is not
present around the trophozoites of piroplasms. High resolution electron micrographs of these organisms (reviewed by Vivier *et al.*, 1970) show that the
parasite's bounding membrane is in direct contact with the cytoplasmic matrix
of the host cell. Indeed Vivier and Petitprez (1969) have proposed that this
feature should have taxonomic significance in classifying doubtful cases such as
Anthemosoma garnhami. Does the absence of this membrane mean that piroplasms enter the cell without invaginating the surface membrane or that, having
entered, for some reason the host cell fails to isolate them in parasitophorous
vacuoles? In the absence of a conoid might we not expect the mode of piroplasm
entry to be similar to that of *Plasmodium*? Like certain malaria parasites, certain
babesias (e.g. *Babesia bigemina*) induce host-cell stippling which fluoresces as

host antigen (Ludford, 1969) using the fluorescent antibody technique. In the absence of an outer membrane the provenance of this stippling should prove interesting in comparison with *Plasmodium*. One final question about host stippling—does this parasite antigen represent shed surface membrane comparable to the plasmanemes of trypanosomes? Both malaria parasites (Brown and Brown, 1965) and piroplasms (Phillips, 1971) display antigenic variation comparable to that of trypanosomes discussed above.

It seems premature to speculate on the relation of the malaria parasite surface to antigenic variation and protective immunity of the host against the disease. Brown (1971) has discussed analogy of the malaria-infected red cell to a hapten protein-carrier antigen system, the parasite strain determinant being equivalent to the carrier protein, the variant specific determinants to hapten moieties. The malaria merozoite has a surface coat (Ladda *et al.*, 1969; Howells, 1970) (Plate 5B), merozoites are agglutinated by immune sera, and agglutination is variant-specific. The coat may provide the variant hapten, the surface membrane the strain determinant—we don't yet know!

The bearing of the origin of the parasitophorous vacuole membrane on parasite physiology is of particular interest for malariologists in view of the supposed dependence of malaria parasites on exogenous ATP and Coenzyme A (Trager, 1950).

Under axenic conditions, *Plasmodium lophurae* will undergo schizogony in a red cell extract provided that ATP is added to the medium. The ATP content of duck erythrocytes infected with this parasite is less than that of unparasitized cells (Trager, 1967). In experimental *P. falciparum* infections of non-immune Negroes, Brewer and Powell (1965) found a positive correlation between erythrocyte ATP content and the rate of increase of the parasitaemia once patent. They claimed that cells with low ATP content were less capable of maintaining structural integrity during the schizogonic cycle of the parasite.

Now most cells (and none are better known than red cells) are impermeable to ATP and CoA, as enzymes hydrolyse these compounds at the plasmalemma. Omission of ATP from *in vitro* cultures of free *P. lophurae* inhibited both parasite development and incorporation of proline. Bonkrekic acid ($\sim 0 \cdot 1$ mM) —a specific inhibitor of adenine nucleotide translocation into mitochondria, produced similar inhibition—an effect which could be counteracted by 10—12 mM ATP but not by AMP (Trager, 1971). Malaria parasites, then, may have in their surface membrane ATP carriers similar to those occurring on the inner mitochondrial membrane. But to reach the parasite's surface membrane from the red cell, ATP has to traverse the membrane of the parasitophorous vacuole and it is difficult to see how this can happen if the vacuolar membrane has the same constitution as the red cell surface membrane which as we know hydrolyses ATP! Are we to suppose that the parasite locally inactivates ATPase? Does the marked increase in Na^+ content of both parasitized and unparasitized cells in malaria (Dunn, 1969) indicate damage to the ATPase ion pump of the red cell? Strangely enough, amino acid transport into malaria parasites is not affected by intracellular sodium or potassium ion concentrations (Sherman *et al.*, 1969), and Read (1970) suggested that these organisms lack gradient-coupled transport systems. As the flow of amino acids into erythrocytes is dependent upon a difference in Na^+ concentration inside and outside, the parasite must rely on the red cell to maintain an ionic gradient compatible with glycolysis and protein synthesis.

Since Rudzinska and Trager (1957) first described phagotrophy in *Plasmodium*, the method of bulk feeding on cell contents by haemosporidians and piroplasms has been debated several times (reviewed by Aikawa, 1971). Briefly, malaria parasites ingest the red cell matrix through a cytostome (Aikawa *et al.*, 1966)—a morphologically differentiated structure, present but inactive in the motile stages of the parasite where it was originally described as the micropyle by Garnham and colleagues (1961). From the cytostomal cavity (which is lined by a double membrane) single membrane-bound food vacuoles are pinched off and in these digestion of the red cell matrix takes place (Aikawa and Thompson, 1971). *Plasmodium elongatum* appears to pinch off vacuoles directly from the cytostome to form boluses which fuse with a larger digestive vacuole and this parasite is unusual in that phagotrophy occurs in the exoerythrocytic stages which take place in haemopoietic cells (Aikawa *et al.*, 1967). Many of the double membrane-bound vacuoles containing red cell matrix seen in sections of *Plasmodium* trophozoites are intrusions of host cell cytoplasm resulting from the parasite's amoeboid activities. In the trophozoites of certain piroplasms (*Babesia*, *Anthemosoma*—see Vivier *et al.*, 1970) a morphologically differentiated cytostome is absent and Vivier and Petitprez (1970) have shown by serial sectioning that the red cell matrix intrudes through several orifices to form a continuous canalicular system, from which presumably, food vacuoles are derived. In mammalian malaria parasites there is a possibility that peripheral pinocytosis of host cell matrix occurs (Cox and Vickerman, 1966) as haemozoin crystals—the remains from digestion in these parasites—are found in small marginal vacuoles.

Although a morphological cytostome is present in other intracellular sporozoa there is no good evidence for its role in feeding. Many workers with Coccidia have described deep invaginations of the inner parasite membrane— suggestive of pinocytotic channels (e.g. Hammond *et al.*, 1967; Vetterling *et al.*, 1971), but more conclusive evidence is needed.

Reaction of the host cell to invasion by a sporozoan parasite varies considerably with the parasite. The intruder may become walled off inside the parasitophorous vacuole as in the case of the cystic form of *Toxoplasma gondii* (Jacobs, 1967), or the whole infected cell may become encapsulated as in the megaloschizonts of *Leucocytozoon simondi* (Desser and Fallis, 1967).

In studying intracellular parasites the boundary between host and parasite is often blurred so that one is hard pressed to say where parasite begins and host ends. This problem is well-illustrated by those sporozoa which become surrounded by a capsule e.g. Mieschers tubes of *Sarcocystis* and the oocyst of *Plasmodium,* but best of all by the remarkable organisms of the genus *Amoebophrya* (Cachon and Cachon, 1969). Once thought to be suctorians or mesozoans, they are now regarded as somewhat aberrant dinoflagellates (Tribe Duboscquodina) which spend the first (trophic) part of their life-cycle as intracellular parasites of radiolarians, ciliates and other dinoflagellates, and the second (sporulating) part free-living in the plankton. As parasites they have a central mastigocoel which is lined by a typical alveolate dinoflagellate pellicle and rows of flagella projecting inwards. But on its outside, *Amoebophrya acanthometrae* is separated from its acantharian host by 6 different layers which could belong to either organism—a granular layer, a reticular layer, a canalicular layer, a paracrystalline layer and a basal unit membrane which connects with the parasite's endoplasmic reticulum. To transform from the trophont to the sporont the organism turns itself inside out—so the host becomes prey and is digested in a large central food vacuole—the six-layered interface along with it!

CONCLUSIONS

1   Extracellular protozoa often display a surface coat outside the surface membrane. This coat may function in binding particles for pinocytosis or in attachment as in other cells but it may also provide a means of escaping the host's immune assault. In trypanosomes the coat may be partly an endogenous secretion containing the variant antigens and partly an exogenous accretion of serum proteins which prevent antigen-antibody interactions.

2   Attachment of protozoan parasites to host cells involves adhesion over a 10–20 nm gap. Specialized desmosomal structures are not necessarily present and taxonomically different protozoa may induce different junctional-complex structures in the same host tissue cell.

3   Intracellular protozoa usually gain access to the host cell by either breaking and entering the surface membrane, or invaginating it; both methods appear to be found within the Coccidia. The method of entry may affect the subsequent relationship with the host-cell cytoplasm and the interaction between parasite and host may result in the formation of new interface structures which do not obviously belong to either, so that the host-parasite interface becomes blurred.

ACKNOWLEDGEMENTS

I am most grateful to Drs M. Aikawa, B. E. Brooker, P. A. D'Alesandro I. Desportes, G. Dorey, D. L. Lee, J. Lom, C. Patton, J. Schrevel, W. Trager and J. M. Vetterling for providing me with illustrations, unpublished results or discussion.

REFERENCES

AIKAWA, M. (1967). Ultrastructure of the pellicular complex of *Plasmodium fallax*. *Journal of Cell Biology* **35**: 103

AIKAWA, M. (1971). Fine structure of malaria parasites. *Experimental Parasitology* **30**: 284

AIKAWA, M. and THOMPSON, P. E. (1971). Localization of acid phosphatase activity in *Plasmodium berghei* and *P. gallinaceum*: an electron microscope study. *Journal of Parasitology* **57**: 603

AIKAWA, M., HUFF, C. G. and SPRINZ, H. (1967). Fine structure of the asexual stages of *Plasmodium elongatum*. *Journal of Cell Biology* **34**: 229

AIKAWA, M., COOK, R. T., SAKODA, J. J. and SPRINZ, H. (1969). Fine structure of the erythrocytic stages of *Plasmodium knowlesi*. *Zeitschrift für Zellforschung und mikroscopische Anatomie* **100**: 271

AIKAWA, M., HEPLER, P. K., HUFF, C. G. and SPRINZ, H. (1966). The feeding mechanism of avian malaria parasites. *Journal of Cell Biology* **28**: 355

ALLISON, A. C., DAVIES, P. and DE PETRIS, S. (1971). Role of contractile microfilaments in macrophage movement and endocytosis. *Nature, London* **232**: 153

ALLSOPP, B. A., NJOGU, A. R. and HUMPHRYES, K. C. (1971). Nature and location of *Trypanosoma brucei* subgroup exoantigen and its relationship to 4S antigen. *Experimental Parasitology* **29**: 271

APFFEL, C. A. and PETERS, J. H. (1970). Regulation of antigenic expression. *Journal of Theoretical Biology* **26**: 47

BEAMS, H. W., KING, R. L., TAHMISIAN, T. N. and DEVINE, R. (1960). Electron microscope studies on *Lophomonas striata* with special reference to the nature and position of the striations. *Journal of Protozoology* **7**: 91

BENNETT, H. S. (1969). In *Handbook of Molecular Cytology*, p. 1261. A. Lima-de-Faria (ed.). Amsterdam: North Holland

BREWER, G. J. and POWELL, R. D. (1965). A study of the relationship between the content of adenosine triphosphate in human red cells and the course of *falciparum* malaria. *Proceedings of the National Academy of Sciences of the United States of America* **54**: 741

BROOKER, B. E. (1970). Desmosomes and hemidesmosomes in the flagellate *Crithidia fasciculata*. *Zeitschrift für Zellforschung und mikroscopische Anatomie* **105**: 155

BROOKER, B. E. (1971a). Flagellar adhesion of *Crithidia fasciculata* to millipore filters. *Protoplasma* **72**: 19

BROOKER, B. E. (1971b). The fine structure of *Crithidia fasciculata* with special reference to the organelles involved in the ingestion and digestion of protein. *Zeitschrift für Zellforschung und mikroscopische Anatomie* **116**: 532

BROWN, K. N. (1971). Protective immunity to malaria parasites provides a model for the survival of cells in an immunologically hostile environment. *Nature, London* **230**: 163

BROWN, K. N. and BROWN, I. N. (1965). Immunity to malaria: antigenic variation in chronic infections of *Plasmodium knowlesi*. *Nature, London* **208**: 1286

CACHON, J. and CACHON, M. (1969). Ultrastructures des *Amoebophryidae* (Peridiniens Dubosquodinida). I. Manifestations des rapports entre l'hôte et le parasite. *Protistologica* **5**: 535

CAMPBELL, M. J. and CAMPBELL, J. H. (1971). Origin and continuity of desmosomes. In *Origin and Continuity of Cell Organelles*, p. 261 J. Reinert and H. Ursprung (eds). Berlin: Springer Verlag

CLARKSON, M. J. and AWAN, M. A. Q. (1968). Studies of antigenic variants of *Trypanosoma vivax*. *Transactions of the Royal Society of tropical Medicine and Hygiene* **62**: 127

CLEVELAND, L. R. and GRIMSTONE, A. V. (1964). The fine structure of the flagellate *Mixotricha paradoxa* and its associated micro-organisms. *Proceedings of the Royal Society, Series B* **159**: 668

COOK, G. M. W. (1968). Glycoproteins in membranes. *Biological Reviews* **43**: 363

COX, F. E. G. and VICKERMAN, K. (1966). Pinocytosis in *Plasmodium vinckei*. *Annals of tropical Medicine and Parasitology*. **60**: 293

CUNNINGHAM, M. P. (1966). The preservation of viable metacyclic forms of *Trypanosoma rhodesiense* and some studies of the antigenicity of the organisms. *Transactions of the Royal Society of tropical Medicine and Hygiene* **60**: 126

CURTIS, A. S. G (1967). *The Cell Surface: its Molecular Role in Morphogenesis*. London: Logos Press

D'ALESANDRO, P. A. (1966). Immunological and biochemical studies of ablastin, the reproduction-inhibiting antibody to *Trypanosoma lewisi*. *Annals of the New York Academy of Science* **129**: 834

D'ALESANDRO, P. A. (1970). Exoantigens of *Trypanosoma lewisi*. *Journal of Parasitology* **56** (Suppt): II (i) p. 65

DESOWITZ, R. S. (1970). Anti-parasitic mechanisms in parasitic infections. *Journal of Parasitology* **56**: 521

DESOWITZ, R. S. and WATSON, H. J. C. (1953). Studies on *Trypanosoma vivax*. VI. The occurrence of antibodies in the sera of infected sheep and white rats, and their influence on the course of infection in white rats. *Annals of tropical Medicine and Parasitology* **47**: 247

DESPORTES, I. (1969). Ultrastructure et développment des grégarines du genre *Stylocephalus*. *Annales des Sciences Naturelles* (*Zoologie*) **11**: 31

DESSER, S. S. and FALLIS, A. M. (1967). The cytological development and encapsulation of megaloschizonts of *Leucocytozoon simondii*. *Canadian Journal of Zoology* **45**: 1061

DUNN, M. J. (1969). Alterations of red blood cell metabolism in simian malaria: evidence for abnormalities of non-parasitized cells. *Military Medicine* **134**: 122

EATON, R. D. P., MEEROVITCH, E. and COSTERTON, J. W. (1970). The functional morphology of pathogenicity in *Entamoeba histolytica*. *Annals of tropical Medicine and Parasitology* **64**: 299

EL-HASHAMI, W. and PITTMAN, F. (1970). Ultrastructure of *Entamoeba histolytica* trophozoites obtained from the colon and from *in vitro* cultures. *American Journal of tropical Medicine and Hygiene* **19**: 215

FRIEND, D. S. (1966). The fine structure of *Giardia muris*. *Journal of Cell Biology* **29**: 317

GARNHAM, P. C. C. (1966). Locomotion in the parasitic protozoa. *Biological Reviews* **41**: 561

GARNHAM, P. C. C., BIRD, R. G., BAKER, J. R. and BRAY, R. S. (1961). Electron microscope studies of motile stages of malaria parasites. II. The fine structure of the sporozoites of *Laverania* (= *Plasmodium*) *falcipara*. *Transactions of the Royal Society of tropical Medicine and Hygiene* **55**: 98

GRAY, A. R. (1965). Antigenic variation in a strain of *Trypanosoma brucei* transmitted by *Glossina morsitans* and *G. palpalis*. *Journal of General Microbiology* **41**: 195

GRAY, A. R. (1967). Some principles of the immunology of trypanosomiasis. *Bulletin of the World Health Organization* **37**: 177

GREENBLATT, C. L., JORI, L. A. and CAHNMANN, H. J. (1969). Chromatographic separation of a rat serum growth factor required by *Trypanosoma lewisi*. *Experimental Parasitology* **24**: 228

GRIFFIN, J. L. and JUNIPER, K. (1971). Ultrastructure of *Entamoeba histolytica* from human amoebic dysentery. *Archives of Pathology* **91**: 271

GRIMSTONE, A. V. (1961). The fine structure of *Streblomastix strix*. Abstr. 1st *International Conference of Protozoology*, Prague: p. 121

HAMMOND, D. M., SCHOLTYSECK, E. and CHABOTAR, B. (1967). Fine structures associated with nutrition of the intracellular parasite *Eimeria auburnensis*. *Journal of Protozoology* **14**: 678

HERBERT, W. J. and MACADAM, R. F. (1971). The immunization of mice with trypanosome plasmanemes (filopodia). *Transactions of the Royal Society of tropical Medicine and Hygiene* **65**: 240

HESSE, E. (1909). Contribution a l'étude des Monocystidées des Oligochaetes. *Archives de zoologie expérimentale et générale* **43**: 27

HOLLINGSHEAD, S., PETHICA, B. A. and RYLEY, J. F. (1963). The electrophoretic behaviour of some trypanosomes. *Biochemical Journal* **89**: 123

HOWELLS, R. E. (1970). Electron microscope observations on the development and schizogony of the erythrocytic stages of *Plasmodium berghei*. *Annals of tropical Medicine and Parasitology* **64**: 305

JACOBS, L. (1967). *Toxoplasma* and Toxoplasmosis. In *Advances in Parasitology* **5**: 1

JADIN, J. and CREEMERS, J. (1968). Ultrastructure et biologie des toxoplasmes. III. Observations de toxoplasmes intraerythrocytaires chez un mammifere. *Acta tropica* **25**: 267

JARUMILINTA, R. and KRADOLFER, F. (1964). The toxic effect of *Entamoeba histolytica* on leucocytes. *Annals of tropical Medicine and Parasitology* **58**: 375

KELLY, D. E. and LUFT, J. H. (1966). Fine structure, development and classification of desmosomes and related attachment mechanisms. In *Electron Microscopy*, Vol. II, p. 401. R. Uyeda (ed.). Tokyo: Maruzen Co. Ltd

KETTERIDGE, D. S. (1971). Studies on rodent-adapted *Trypanosoma vivax*. Ph.D. Thesis. Glasgow University

LADDA, R. L., AIKAWA, M. and SPRINZ, H. (1969). Penetration of erythrocytes by merozoites of mammalian and avian malaria parasites. *Journal of Parasitology* **55**: 633

LANHAM, S. M. (1968). Separation of trypanosomes from the blood of infected rats and mice by anion exchangers. *Nature, London* **218**: 1273

LEE, D. L. (1969). The structure and development of *Histomonas meleagridis* (Mastigamoebidae: Protozoa) in the female reproductive tract of its intermediate host *Heterakis gallinarum* (Nematoda). *Parasitology* **59**: 877

LEE, D. L., LONG, P. L., MILLARD, B. J. and BRADLEY, J. (1969). The fine structure and method of feeding of the tissue parasitizing stages of *Histomonas meleagridis*. *Parasitology* **59**: 171

LINCICOME, D. R. (1958). Growth of *Trypanosoma lewisi* in the heterologous mouse host. *Experimental Parasitology* **7**: 1

LOM, J. (1971). *Trichophrya piscium* Bütschli: a pathogen or an ectocommensal? An ultrastructural study. *Folia Parasitologia* (Prague) **18**: (in press)

LOM, J. and CORLISS, J. O. (1970). Attachment structures in ectoparasitic protozoa of fishes and their possible relation to pathogenicity. *Journal of Parasitology* **56** (Suppt) II (i): 212

LUDFORD, C. G. (1969). Fluorescent antibody staining of four *Babesia* species. *Experimental Parasitology* **24**: 327

MACGREGOR, H. C. and THOMASSON, P. A. (1965). The fine structure of two archigregarines, *Selenidium fallax* and *Ditrypanocystis cirratuli*. *Journal of Protozoology* **12**: 438

McLAREN, D. J. and PAGET, G. E. (1968). A fine structural study of the merozoite of *Eimeria tenella* with special reference to the conoid apparatus. *Parasitology* **58**: 561

MACMILLAN, W. G. (1969). Some aspects of the biology of *Nematocystis magna* Schmidt. Ph.D. Thesis. Glasgow University

MILLER, H. C. and TWOHY, D. W. (1967). Infection of macrophages in culture by leptomonads of *Leishmania donovani*. *Journal of Protozoology* **14**: 781

MOLYNEUX, D. H. (1969). The fine structure of the epimastigote forms of *Trypanosoma lewisi* in the rectum of the flea, *Nosopsyllus fasciatus*. *Parasitology* **59**: 55

NIELSEN, M. H. (1970). Phagocytosis by *Trichomonas vaginalis* Donne. In *Microscopie Electronique*, Vol. III: 389. P. Favard (ed.). Paris

ORMIÈRES, R. and DAUMAL, J. (1970). Étude ultrastructurale de la partie antérieure d'*Epicavus araeoceri* O. and D. *Protistologica* **6**: 97

PHILLIPS, R. S. (1971). Antigenic variation in *Babesia rodhaini* demonstrated by immunization with irradiated parasites. *Parasitology* **63**: 315

PORCHET-HENNERÉ, E. and VIVIER, E. (1971). Ultrastructure comparée des germes infectieux (sporozoites, merozoites, schizozoites, endozoites, etc.) chez les sporozoaires. *L'Année Biologique* **10**: 77

PRATHAP, K. and GILMAN, R. (1970). The histopathology of acute intestinal amoebiasis. A rectal biopsy study. *American Journal of Pathology* **60**: 229

READ, C. P. (1970). Some physiological and biochemical aspects of host-parasite relations. *Journal of Parasitology* **56**: 643

ROBERTS, W. L. and HAMMOND, D. M. (1970). Ultrastructural and cytological studies of sporozoites of four *Eimeria* species. *Journal of Protozoology* **17**: 76

ROBERTS, W. L., HAMMOND, D. M. and SPEER, C. A. (1970). Ultrastructural study of the intra- and extracellular sporozoites of *Eimeria callospermophili* and *E. larimerensis*. *Journal of Parasitology* **56**: 918

RUDZINSKA, M. A. (1969). The fine structure of malaria parasites. *International Review of Cytology* **25**: 161

RUDZINSKA, M. A. and TRAGER, W. (1957). Intracellular phagotrophy by malaria parasites: an electron microscope study of *Plasmodium lophurae*. *Journal of Protozoology* **4**: 190

RUDZINSKA, M. A. and TRAGER, W. (1968). The fine structure of trophozoites and gametocytes of *Plasmodium coatneyi*. *Journal of Protozoology* **15**: 73

RUDZINSKA, M. A., D'ALESANDRO, P. A. and TRAGER, W. (1964). The fine structure of *Leishmania donovani* and the role of the kinetoplast in the leishmania-leptomonad transformation. *Journal of Protozoology* **11**: 166

RYLEY, J. F. (1969). Ultrastructural studies on the sporozoite of *Eimeria tenella*. *Parasitology* **59**: 67

SCHOLTYSECK, E. (1969). Electron microscope studies of the effect upon the host cell of various developmental stages of *Eimeria tenella* in the natural chicken host and in tissue cultures. *Acta veterinaria* **38**: 153

SCHOLTYSECK, E., MEHLHORN, H. and FRIEDHOFF, K. (1970). The fine structure of the conoid of Sporozoa and related organisms. *Zeitschrift für Parasitenkunde* **34**: 68

SCHREVEL, J. (1968). L'ultrastructure de la région antérieure de la grégarine Selenidium et son intérêt pour l'étude de la nutrition chez les sporozoaires. *Journal de Microscopie* **7**: 391

SCHREVEL, J. and VIVIER, E. (1966). Étude de l'ultrastructure et du rôle de la région antérieure de gregarinas parasites d'annélides polychaetes. *Protistologica* **2**: 17

SEED, J. R. (1964). Antigenic similarity among culture forms of the 'brucei' group of trypanosomes. *Parasitology* **54**: 593

SHEFFIELD, H. G., GARNHAM, P. C. C. and SHIROSIMI, T. (1971). The fine structure of the sporozoite of *Lankesteria culicis. Journal of Protozoology* **18**: 98

SHERMAN, I. W., RUBLE, J. A. and TANIGOSHI, L. (1969). Incorporation of C^{14}-labelled amino acids by malaria (*Plasmodium lophurae*). I. Role of ions and amino acids in the medium. *Military Medicine* **134**: 954

SOLOVJEV, M. M. and PLATOVA, G. D. (1969). Biology of *Lamblia* in connexion with membranous digestion in small intestine. In *Progress in Protozoology*, p. 317. Leningrad: Academy of Science of the U.S.S.R.

SPRAGUE, V. and VERNICK, S. H. (1968). Light and electron microscope study of a new species of *Glugea* (Microsporidia: Nosematidae) in the 4-spined stickleback *Apeltes quadracus. Journal of Protozoology* **15**: 547

STRAUSS, P. R. (1971). The effect of homologous rabbit antiserum on the growth of *Leishmania tarentolae*—a fine structure study. *Journal of Protozoology* **18**: 147

TOBIE, J. E. and COATNEY, G. R. (1961). Fluorescent antibody staining of human malaria parasites. *Experimental Parasitology* **11**: 128

TRAGER, W. (1950). Studies on the extracellular cultivation of an intracellular parasite (avian malaria). I. Development of the organisms in erythrocyte extracts and the favouring effect of adenosine triphosphate. *Journal of Experimental Medicine* **92**: 349

TRAGER, W. (1967). Adenosine triphosphate and the pyruvic and phosphoglyceric kinases of the malaria parasite *Plasmodium lophurae. Journal of Protozoology* **14**: 110

TRAGER, W. (1972). Effects of Bongkrekic acid on malaria parasites (*Plasmodium lophurae*) developing extracellularly *in vitro*. In *The Comparative Biochemistry of Parasites*. Janssen Pharmaceutica (in press)

TUZET, O. and GALANGAU, V. (1968). Ultrastructure des rhizoides de la grégarine *Grebnickiella gracilis* Bathia. *Compte rendu Hebdomadaire des séances de l'Académie des sciences* **266**: 1401

VAVRA, J. (1965). Etude au microscope électronique de la morphologie et du développment de quelques microsporidies. *Compte rendu Hebdomadaire des séances de l'Académie des sciences* **261**: 3467

VAVRA, J. and SMALL, E. B. (1969). Scanning electron microscopy of gregarines (Protozoa, Sporozoa) and its contribution to the theory of gregarine movement. *Journal of Protozoology* **16**: 745

VETTERLING, J. M., TAKEUCHI, A. and MADDEN, P. A. (1971). Ultrastructure of *Cryptosporidium wrairii* from the guinea pig. *Journal of Protozoology* **18**: 248

VICKERMAN, K. (1969a). The fine structure of *Trypanosoma congolense* in its blood stream phase. *Journal of Protozoology* **16**: 54

VICKERMAN, K. (1969b). On the surface coat and flagellar adhesion in trypanosomes. *Journal of Cell Science* **5**: 163

VICKERMAN, K., and LUCKINS, A. G. (1969). Localization of variable antigens in the surface coat of *Trypanosoma brucei* using ferritin-conjugated antibody. *Nature, London* **224**: 1125

VICKERMAN, K. (1971). Morphological and physiological considerations of extracellular blood protozoa. In *Ecology and Physiology of Parasites*, p. 58. A. M. Fallis (ed.). Toronto University Press

VIVIER, E. and PETITPREZ, A. (1969). Observations ultrastructurales sur l'hématozoaire *Anthemosoma garnhami* et examen des critères morphologiques utilisables dans la taxonomie chez les Sporozoaires. *Protistologica* **5**: 363

VIVIER, E. and PETITPREZ, A. (1970). Étude par la technique des coupes seriées du système vacuolaire chez un haematozoaire, *Anthemosoma garnhami*. In *Microscopie Electronique*, Vol. III, p. 413. P. Favard (ed.). Paris

VIVIER, E., DEVAUCHELLE, G., PETITPREZ, A., PORCHET-HENNERÉ, E., PRENZIER, J., SCHREVEL, J. and VINCKIER, D. (1970). Observations de cytologie comparée chex les sporozoaires. I. Les structures superficielles chez les formes végétatives. *Protistologica* 6: 127

VOLLER, A. (1965). Immunofluorescence and humoural immunity to *Plasmodium berghei*. *Annales de la Société belge de Médecine tropicale* 45: 385

WATKINS, J. F. (1964). Observations on antigenic variation in a strain of *Trypanosoma brucei* growing in mice. *Journal of Hygiene* 62: 69

WEITZ, B. G. F. (1960a). A soluble protective antigen of *Trypanosoma brucei*. *Nature, London* 185: 788

WEITZ, B. G. F. (1960b). The properties of some antigens of *Trypanosoma brucei*. *Journal of General Microbiology* 23: 589

WILLIAMSON, J. and BROWN, K. N. (1964). The chemical composition of trypanosomes. III. Antigenic constituents of *Brucei* trypanosomes. *Experimental Parasitology* 15: 44

WRIGHT, K. A. and HALES, H. (1970). Cytochemistry of the pellicle of bloodstream forms of *Trypanosoma* (*Trypanozoon*) *brucei*. *Journal of Parasitology* 56: 671

WRIGHT, K. A., LUMSDEN, W. H. R. and HALES, H. (1970). The formation of filopodium-like processes by *Trypanosoma* (*Trypanozoon*) *brucei*. *Journal of Cell Science* 6: 285

CELL MEMBRANES AND IMMUNE RESPONSES

A. C. ALLISON

Clinical Research Centre, Harrow, Middlesex

Reciprocal interactions of the plasma membranes of mammalian, parasitic and microbial cells with the immune system are complex. Blood and tissue group specificity, which are important for transfusion and organ transplantation, depend on membrane antigens, and many immune responses against parasites and pathogenic microorganisms are elicited by membrane constituents. The immunogenic groups of membranes include carbohydrates, proteins and complex lipids. These can stimulate the formation of several classes of antibodies and also, in many cases, cell-mediated immunity. Antibodies vary in the effects they produce on membranes of target cells. They can agglutinate or lyse cells, opsonize them for phagocytosis, or actually induce endocytosis or exocytosis.

Cell-mediated immunity involves lymphocytes that can recognize and interact specifically with target cell antigens; as a result of the interaction, macrophages are concentrated in the reaction site. The macrophages may be 'activated', showing an increased content of lysosomal enzymes and capacity to kill ingested organisms or cells with which they come into contact. Antibody can act synergistically with cell-mediated immunity, for example when cytophilic antibody and activated macrophages together destroy parasites or foreign cells. On the other hand, certain types of antibody can block effects of cell-mediated immunity. Only a few aspects of these interactions, which reflect my personal interests, can be considered in this paper. I shall begin by reminding you of some features of membrane structure which are relevant to immune responses against cells and parasites.

THE PLASMA MEMBRANE OF MAMMALIAN CELLS

Various methods have been developed for the isolation of plasma membranes, and it has been possible to show that their compositions are somewhat different from those of membranes elsewhere in the cell. Thus, plasma membranes have a

93

relatively high proportion of sphingomyelin and cholesterol. Cholesterol 'stiffens' membranes, and the plasma membrane is less flexible than other membranes (Bosmann *et al.*, 1969). About 30% of the plasma membrane mass is made up of proteins. According to the classical Danielli-Davson model, the membrane consists primarily of a bilayer of lipid with hydrophobic groups facing inward and hydrophilic groups facing outwards and interacting with protein. In electron micrographs of suitably fixed material, a characteristic trilaminar membrane structure is observed. Recent evidence suggests that this model is true for most of the membrane structure, but that in some regions proteins with hydrophobic side chains extend into the bilayer and interact with non-polar portions of lipid molecules. Many of the functions of membranes, including selective transport across them, are thought to be due to protein carriers or enzymes within the membrane itself.

Many membrane proteins have carbohydrate moieties, and the carbohydrate groups of the outer surface make an important contribution to the antigenicity of cells. The major blood group antigens, for example, are the terminal disaccharides of carbohydrate side chains which are attached to a peptide backbone. In electron micrographs the outer layer of membrane is seen to be covered by a glycopeptide cell coat, the thickness of which varies. The coat can readily be sheared from the plasma membrane and remain attached to substrates.

LYSIS OF CELLS BY ANTIBODY AND COMPLEMENT

Among the most remarkable effects on plasma membranes is the capacity of antibody and complement to bring about lysis. A single molecule of IgM attached to a cell is sufficient for lysis, but in the case of IgG antibody more than one molecule is required; apparently two IgG molecules on the cell membrane must be sufficiently close together for complement components to form a bridge between them. Green *et al.* (1959) showed that after combination of antibody and complement with cells, low-molecular weight cytoplasmic contents such as potassium and free amino acids leak out, but that macromolecules such as protein and RNA are lost only if osmotic effects are not counterbalanced by the presence of protein in the medium. Proteins are trapped within the plasma membrane, and an energy-driven pump offsets cation leakage inwards. If the cation permeability of such a system increases beyond the capability of the pump, the non-diffusable proteins within will effect a Donnan redistribution of ions, and the concomitant rise in osmotic pressure and swelling will rupture the membrane and release its contents. Green and his colleagues concluded that antibody and complement produce functional holes in the cell membrane, large enough to permit the free exchange of water and ions such as K^+ and Na^+ but too small to allow passage of macromolecules.

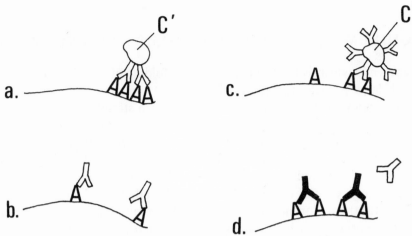

FIG. 1.   Cytotoxic and blocking effects of antibody. a. When antigenic sites on a cell membrane are close together, two molecules of IgG can form a bridge, fix complement and initiate lysis. b. When antigenic sites are far apart, IgG molecules become attached, but bridging is not possible. Hence there is no lysis, but antigenic sites are blocked so that IgM antibodies cannot lyse cells. c. In the absence of IgG attachment, a single molecule of IgM can become attached to an antigenic site, fix complement and lyse cells. d. Non-complement-fixing antibody (e.g. IgA, IgG_2 or IgG_4) can become attached to antigenic sites and block lysis by subclasses of antibody which can fix complement.

These holes have been visualized in electron micrographs of negatively stained membranes (Humphrey and Dourmashkin, 1969). When the surface of the membrane lies flat on the electron microscope grid, a dark central part of the hole is observed, surrounded by a lighter ring. When a folded edge of the membrane is present, the clear ring can be seen projecting from the membrane surface, forming a hollow cylinder filled with negative stain. The complement holes are thought to be due to a rearrangement occurring in the membrane to form micellar structures in a predominantly lipid surface layer. These holes are observed in the walls of bacteria, and formation of such holes, along with the action of the enzyme lysozyme (muramidase), contributes to bacterial lysis and host protection.

For complement-dependent lysis by IgG antibody the antigenic sites on the target cells must be close together to allow 'bridging'. If the antigenic sites are widely separated, IgM antibodies can bring about lysis (since only a single molecule with an antigenic site is required), but this can be prevented by prior attachment of IgG molecules to the antigenic sites. Likewise, combination with non-complement-fixing antibodies (such as the IgA class or human IgG_2 or IgG_4 subclasses) will prevent subsequent antibody and complement lysis irrespective of the distribution of the antigenic sites (Fig. 1).

REQUIREMENTS FOR ENDOCYTOSIS AND EXOCYTOSIS

The term endocytosis is used to describe the infolding of the plasma membrane to engulf large particles (phagocytosis) or to take in smaller amounts of fluid medium (pinocytosis). Several factors in serum can stimulate endocytosis. The most potent is anticellular antibody, including a macroglobulin in serum which reacts with mouse macrophages and has the properties of an interspecies antibody (Cohn and Parks, 1967). Stimulated endocytosis is followed by increased formation of lysosomal hydrolases. It has long been known that antigen-antibody complexes near equivalence are rapidly endocytosed into macrophages and degraded there (Sorkin and Boyden, 1959). Immunoglobulins G and M are attached to specific and independent sites on macrophages (Lay and Nussenzweig, 1968; Huber et al., 1971), but the attachment of native immunoglobulins does not initiate endocytosis. In contrast, when the configuration of immunoglobulin is altered by heating so that it can fix complement in the absence of antigen, it is rapidly taken into macrophages (Hess and Luscher, 1970); complement is not necessary for this process. It therefore seems likely that the natural factor stimulating endocytosis is immunoglobulin which has undergone a configurational change in the course of interaction with antigen (which may be the membrane of the cell itself). This would have obvious advantages as a trigger mechanism, since it would ensure that immunoglobulins present in tissue fluids would not stimulate endocytosis unless they had already reacted with antigen. A group of immunoglobulins termed cytophilic antibodies, described below, has relatively high affinity for macrophages even when not combined with antigen, but endocytosis again appears to follow interaction with antigen.

We have recently presented evidence (Allison et al., 1971) that endocytosis is brought about by contraction of a network of actomyosin-like microfilaments beneath the plasma membrane. This system is selectively paralyzed by macrolide fungal products known as cytochalasins, which reversibly block phagocytosis of bacteria and pinocytosis of colloidal gold.

The reverse process takes place when, during the course of acute hypersensitivity reactions (e.g. in asthma and hay fever), granules containing histamine, serotonin and slow-reacting substance are released from mast cells. The attachment of IgE and related antibodies to leucocytes does not trigger granule release, but this follows subsequent combination with antigen. Again, cytochalasins strongly inhibit the release, which appears to depend on contraction of the actomyosin-like microfilament system in the peripheral cytoplasm. Calcium ions from the external medium appear to be required for endocytosis and exocytosis (Rabinovitch, 1967; Rubin, 1970) and are the universal trigger mechanism for activation of actomyosin-like contractile systems (Ebashi and Endo, 1968). It therefore seems likely that specific stimuli (attachment of the right kind of antibody and antigen to the surface of macrophages, polymorphs or mast cells) increases calcium permeability through the cell membrane and so brings about contraction of the peripheral cytoplasmic microfilament system. This can result in endocytosis of particles and medium or in exocytosis of granules.

CELL-MEDIATED IMMUNITY AND MACROPHAGE ACTIVATION

It is now widely accepted that thymus-derived (T) lymphocytes play a major role in cell-mediated immunity. On exposure to antigen, T cells in the 'thymus-dependent' areas of spleen and lymph nodes undergo transformation into blast cells, proliferate and are released through the lymph to the blood stream. They circulate as small or medium-sized long-lived lymphocytes and enter the site where antigen is present. There they again undergo transformation into blasts, acquire an increased concentration of lysosomal hydrolytic enzymes and liberate various biological mediators. One of these (migration inhibitory factor, MIF) immobilizes macrophages at the reaction site. The macrophages in an animal undergoing a cell-mediated immune response become 'activated': they have a raised content of lysosomal hydrolases and show an increased capacity to kill ingested organisms (Mackaness, 1970). Although the stimulus to macrophage activation is antigen-specific, the effect is not. In other words, if macrophages have been activated as a result of inoculation with tubercle bacilli, they also kill with great efficiency unrelated bacteria such as *Listeria*, as well as intracellular protozoa such as *Toxoplasma* and viruses. As discussed further below, activated macrophages also show increased capacity to kill tumour cells. Hence this manifestation of lymphocyte-mediated immunity is termed non-specific cellular immunity.

CYTOPHILIC ANTIBODY

Macrophages may, however, acquire capacity to react specifically with antigen after uptake of cytophilic antibodies. These were defined by Boyden (1964) as globulin components of antiserum which become attached to certain cells in such a way that these cells are capable of specifically absorbing antigen. Cytophilic antibodies are usually demonstrated by a rosette test, in which peritoneal macrophages are incubated with serum, washed, and shown to have acquired the capacity to bind antigen, often in the form of cells or attached to cells. The formation and properties of cytophilic antibodies have been reviewed by Nelson (1969). Since they may be of importance in immunity against parasites and tumours, cytophilic antibodies are discussed at some length in this paper.

Sheep erythrocytes have been widely used for studies of cytophilic anti-bodies, both as antigens and as carriers of antigens. Cytophilic antibodies are unusual in that they are by no means invariably produced following the usual forms of antigenic stimulation and their presence is not consistently related to that of other types of antibodies. As a rule, inoculation of cells or soluble antigens in Freund's complete adjuvant is more likely to stimulate formation of cytophilic antibodies than in the absence of the complete adjuvant. Under certain circumstances inoculation of foreign cells intravenously or by other routes in the absence of adjuvant gives rise to cytophilic antibodies. The production in C57BL/6J mice of cytophilic antibodies directed against histo-compatibility antigens (H2) of A/J mice, using the A/J tumour sarcoma 1 as a test cell for attachment to macrophages, has been described by Nelson (1969). The cytophilic alloantibodies were produced after grafts of either A/J skin or sarcoma 1 cells themselves.

In guinea pigs cytophilic antibody is found in the slow $7S$ γ_2-globulin fraction (Nelson, 1969); in mice it is usually but not always $7S$ γ_2. Capacity to attach to macrophages depends on the presence of the Fc fragment. Treatment with mercaptoethanol followed by iodoacetamide abolishes cytophilic activity. Cytophilic antibodies attached to the macrophage surface are in equilibrium with cytophilic antibodies in the surrounding medium, but the affinity of cell for antibody is relatively high. Cytophilic antibody against one antigen can be displaced from a macrophage surface by cytophilic antibody against a second antigen (Berken and Benacerraf, 1966), and incubation with normal homologous sera often displaces some cytophilic antibody from macrophages. Frequently normal sera also have cytophilic antibodies against heterologous tumour and other cells in low titre.

KILLING OF TARGET CELLS BY LYMPHOCYTES

Immunity against tumours, like that against homografts, can be transferred from one animal to a syngeneic recipient by lymphoid cells but usually not by serum (Boyse *et al.*, 1969; Law, 1966; Allison, 1971). It has therefore been concluded that, as a rule, immunity against tumours is cell-mediated and that lymphocytes themselves play an important role in killing tumour cells. Observa-tions of specific killing by sensitized lymphoid cells of tumour cells *in vitro* (Hellström and Hellström, 1969) have been taken in support of this interpreta-tion. Cerottini *et al.* (1970) have shown that pretreatment with anti-θ antibody reduces the capacity of sensitized mouse cells to kill tumour cells, and conclude that in their system thymus-dependent (T) lymphocytes exert direct cytotoxic effects on tumour cells. However, in other systems target cells sensitized by antibody may be killed by populations of spleen cells depleted of T lymphocytes (see Harding *et al.*, 1971). It is supposed that B (bone-marrow derived) lymphocytes are responsible both for the production of the antibody which is attached to target cells (although a helper effect of T cells may be required) and also for cytotoxic effects. However, the possibility that macrophages or granulocytes may make an important contribution to the cytotoxicity of spleen cell suspensions for antibody-coated target cells remains open, and would be consistent with other observations discussed below.

BLOCKING AND UNBLOCKING ANTIBODIES

Experimental animals and human patients often have lymphoid cells which are able to kill their own tumour cells in culture, but at the same time they have in their serum factors which can specifically block this cytotoxicity (Hellström and Hellström, 1969). The blocking factors appear to be specific antibodies or antigen-antibody complexes. The tumours are thought to be able to grow because an effective immune response is blocked: in other words, growth is 'enhanced' by specific antibodies or immune complexes.

Recently yet another factor or group of factors has been discovered in the sera of animals after tumour regression. These are termed 'unblocking' factors because they can reverse the inhibition by blocking factors of specific cell-mediated cytotoxicity in culture. Moreover, such unblocking sera, injected into mice and rats, can bring about regressions of Moloney sarcomas and polyoma tumours (Hellström *et al.*, 1969; Hellström and Hellström, 1970; Bansal and Sjögren, 1971).

CELL KILLING BY ACTIVATED MACROPHAGES

Gorer (1956) summarized observations that macrophage reactions are prominent in rejection of tumour cells, especially during the phase before extensive vascularization. Many investigators have found that immunity against tumours can be transferred more efficiently by peritoneal exudate cells (rich in macrophages) than lymph node or spleen cells (see Nelson, 1969); and Amos (1962) and Bennett (1965) have shown that peritoneal macrophages alone (separated magnetically after ingestion of iron oxide or after growth on collagen) were effective in transferring immunity against tumour cells. Interactions *in vitro* between macrophages from immunized mice and tumour cells (sarcoma 1) have been investigated by Granger and Weiser (1964, 1966). The tumour cells were killed in a manner that was immunologically specific, involved adherence of the macrophages to the target cells but not phagocytosis, and required the metabolic integrity of the macrophages.

Evans and Alexander (1970) analysed the immunity induced in DBA/2 mice by irradiated syngeneic L51784 lymphoma cells. The resistance could be transferred to syngeneic untreated mice by spleen or lymph-node cells, but neither the serum nor spleen cells of immune mice were cytotoxic or growth inhibitory to tumour cells *in vitro*. However, cultures of peritoneal exudate cells from immune mice were found to exert an immunologically specific cytotoxic effect on the tumour cells. Syngeneic macrophages from untreated mice after incubation with spleen cells from immune animals acquired the capacity to kill tumour cells. In later experiments Alexander and Evans (1971) found that treatment of mouse peritoneal macrophages in culture with bacterial endotoxin or double-stranded polyribonucleotides resulted in 'activation', as judged by the number and size of their lysosomes. Prolonged contact of lymphoma cells with the activated macrophages, but not with normal macrophages, resulted in marked inhibition of the capacity of tumour cells to grow when transferred to fresh medium. Macrophages taken from animals inoculated with endotoxin or double-stranded polyribonucleotides likewise showed much greater cytotoxicity for lymphoma cells than did macrophages from untreated animals.

Other recent observations on the relationship between parasitic or protozoal infection and tumour growth also suggest that activated macrophages play an important part in antitumour immunity. Weatherly (1970) reported that the incidence of naturally occurring mammary tumours in mice was reduced when they were infected with the nematode *Trichinella spiralis*. Keller *et al.* (1971) found that the growth of Walker sarcoma was suppressed in rats recently infected with *Nippostrongylus brasiliensis* but enhanced in rats infected for 10 or 30 days with the parasite. The enhancing effect could be transferred with serum from animals immune to the parasite. Growth of a syngeneic adenocarcinoma in mice was suppressed in mice infected with *N. brasiliensis*, irrespective of the timing of tumour and parasite inocula, and the suppression was overcome by antilymphocyte serum. Analysing the mechanism by which *N. brasiliensis* infection increases immunity against tumour cells in rats, Keller and Jones (1971) found that peritoneal exudate macrophages 'activated' by infection with the parasite, or simply by inoculation of proteose peptone, could transfer to syngeneic rats resistance against the tumour. Normal macrophages did not have this effect, and only the activated macrophages were able rapidly to ingest or destroy tumour cells in culture. The IgG_2 fraction from serum of animals immune to *N. brasiliensis* abolished the tumour inhibitory effect of the activated macrophages *in vitro*.

Working independently at Stanford, Hibbs and his colleages (1971) have found that infections of mice by protozoan parasites increase resistance to tumours. Infections with *Toxoplasma gondii* and *Besnoitia jellisoni* persist indefinitely in mice and give prolonged macrophage activation with increased capacity to kill intracellular bacteria such as *Listeria* and viruses such as Mengovirus. Hibbs *et al.* (1971) found significant resistance against the development of spontaneous mammary tumours in infected C_3H/He mice: the tumour incidence at 6 months was 62% in controls, as compared with 26% in those infected with *Toxoplasma* or *Besnoitia*. The incidence of spontaneously occurring leukaemia in AKR mice at 6 months was 69%, that in *Toxoplasma*-infected mice was 19%, while there were no leukaemias in *Besnoitia*-infected mice. Resistance against Friend virus-induced lymphoreticular malignancy and transplants of sarcoma 180 cells was also observed in mice infected with the protozoa. Hibbs and his colleagues concluded that macrophages might play a major role in this type of tumour resistance, and found (Hibbs *et al.*, 1972) that activated macrophages, taken from the peritoneal cavities of infected animals, kill L cells and transplantable syngeneic (C_3H adenocarcinoma) and allogeneic (BALB/c sarcoma) tumour cells in culture.

SYNTHESIS

At first sight some of the observations just reviewed seem puzzling. That *Nippostrongylus* and other parasites should exert some sort of adjuvant effect on immune responses against tumours is understandable: the helminth infection is known, for example, to increase the production of reaginic antibodies against unrelated antigens (Orr and Blair, 1961). But why should IgG_2 antibody from rats immune to *Nippostrongylus* enhance tumour growth? There is nothing to suggest that the tumour and parasite are antigenically related. How are these observations related to those on blocking and unblocking antibodies?

A hypothesis can be advanced which accommodates the known facts and predicts certain consequences that are testable. My suggestion is that a major killer of tumour cells is the activated macrophage. The results of Granger and Weiser (1964, 1966) Alexander and Evans (1971), Keller and Jones (1971) and of Hibbs *et al.* (1972) leave little room for doubt that macrophages activated in various ways—by specific immunity, immunity to unrelated organisms (*Nippostrongylus, Toxoplasma, Besnoitia*), or following exposure to bacterial endotoxin, double-stranded polyribonucleotides or peptone—kill tumour cells much more efficiently than do untreated macrophages. The effectiveness of transfers of activated macrophages in recipient animals, and of macrophage activation by infections with *Nippostrongylus* or protozoa in inhibiting tumour induction or transplantation, suggests that activated macrophages are able to prevent tumour cell growth *in vivo* as well as *in vitro*. A variety of agents stimulating reticuloendothelial activity inhibit tumour growth (Yashphe, 1971), which is consistent with this interpretation. Inoculation of Freund's adjuvant is a well-known way to obtain activated macrophages, and under certain conditions markedly inhibits tumour cell growth *in vivo*. Likewise, the anti-tumour effects of double-stranded polyribonucleotides, which have recently attracted attention, may be related to their capacity to activate macrophages and, perhaps, act as adjuvants for immune responses. To obtain macrophage activation *in vivo* by immunization with tumour or parasites, cell-mediated immunity would be required (as in Mackaness, 1970). This could explain why transfers of sensitized lymphocytes confer protection and why treatment of mice with anti-lymphocytic serum abolishes the tumour inhibitory effect of *Nippostrongylus* infection (Keller *et al.*, 1971).

Activated macrophages would not show any immunological specificity, whereas there is evidence that macrophage-mediated tumour cytotoxicity in immune animals does manifest such specificity (Granger and Weiser, 1966; Nelson, 1969; Evans and Alexander, 1970). Target cell specificity could be brought about by antibody, either specific cytophilic antibody on the macrophage or target cell, or another type of antibody which after attachment to the target cell is able through configurational changes in its Fc portion to react with receptors on the macrophage membrane (Fig. 2). Either would ensure close and prolonged contact of the activated macrophage and target cell, as a result of which the latter would be killed.

In actively immunized animals, specific cytophilic antibody would probably already be present on macrophages (Nelson, 1969), and tumour cells grown

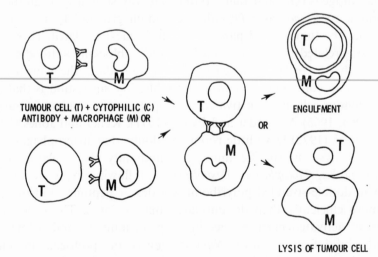

TUMOUR CELL (T) + CYTOPHILIC (C)
ANTIBODY + MACROPHAGE (M) OR

ENGULFMENT

OR

LYSIS OF TUMOUR CELL

FIG. 2. Diagram showing how cytophilic antibody on macrophages, or antibody on tumour cells having an Fc portion with a macrophage receptor site, can bring about close association of activated macrophage and tumour cell, leading to tumour cell lysis or engulfment. The latter is less common and may follow lysis.

in vivo would frequently have antibody on their surfaces. Acquisition by macrophages of specific cytotoxic capacity as a result of incubation with immune lymphocytes (Evans and Alexander, 1970; Harding *et al.*, 1971) could likewise be due to cytophilic antibodies. Low levels of 'natural' cytophilic antibodies occur in many sera (Nelson, 1969). The killing by activated macrophages of rat tumour cells described by Keller and Jones (1971) probably required antibody on the surface of the tumour cells since it was abolished by pretreatment of the target cells with goat antiserum against rat globulins (Keller, personal communication).

The cytotoxic mechanism might be local activation of part of the complement system (of which macrophages possess several components), local exocytosis of lysosomal hydrolases (such as phospholipase A) or some hitherto undisclosed mechanism. 'Unblocking' antibody would according to this hypothesis represent cytophilic or other antibody functioning as a bridge between macrophages and target cells, and so promoting close contact and killing.

In contrast, blocking factors would interfere with this contact of macrophages and target cells. Some subclasses of antibodies (e.g. human IgG_2 and IgG_4) do not react with macrophage receptors (Huber *et al.*, 1971), and could become attached to target cells without attracting the attention of macrophages.

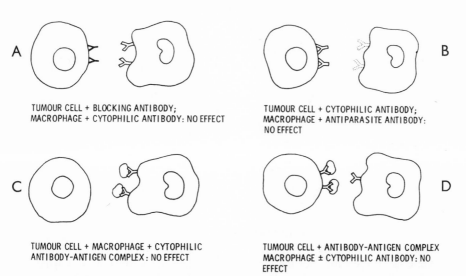

A   TUMOUR CELL + BLOCKING ANTIBODY;
MACROPHAGE + CYTOPHILIC ANTIBODY: NO EFFECT

B   TUMOUR CELL + CYTOPHILIC ANTIBODY;
MACROPHAGE + ANTIPARASITE ANTIBODY:
NO EFFECT

C   TUMOUR CELL + MACROPHAGE + CYTOPHILIC
ANTIBODY-ANTIGEN COMPLEX: NO EFFECT

D   TUMOUR CELL + ANTIBODY-ANTIGEN COMPLEX
MACROPHAGE ± CYTOPHILIC ANTIBODY: NO
EFFECT

FIG. 3. Various mechanisms by which macrophage-mediated cytotoxicity might be blocked. A. Subclasses of antibody on the tumour cell lacking Fc able to combine with macrophage receptors could block effects of macrophages, even if they were armed with antitumour cytophilic antibody. B. The presence on the surface of macrophages of cytophilic antibodies against unrelated antigens (e.g. *Nippostrongylus*) could prevent the macrophages from attacking target cells. C and D. The presence of antigen-antibody complexes on the surface of tumour cells or of macrophages could block the capacity of the macrophages to kill tumour cells.

104 A. C. ALLISON

Likewise antigen-antibody complexes could become attached to macrophages or target cells and prevent their close contact (Fig. 3). Increased levels of cytophilic antibody against an unrelated antigen such as *Nippostrongylus* in the rat would displace low levels of antitumour cytophilic antibodies from the surface of macrophages (Fig. 3) and so 'enhance' tumour growth, as observed by Keller and Jones (1971).

Immunity against tumours is evidently a complex process. Apart from lymphocyte-mediated cytotoxicity, which may play a role (Cerottini *et al.*, 1970), and antibody-mediated lysis of tumour cells in the presence of complement, which operates against some lymphoma cells (Boyse *et al.*, 1969), activated macrophages may represent a major system of defence against tumours, especially in the presence of the right kind of antibody.

The most important prediction that can be made on this hypothesis is that if syngeneic macrophages, activated by some immunologically unrelated stimulus, and 'unblocking' antibody are transferred at the same time to non-immunized recipients they will confer much greater protection against tumours than will either procedure alone. Moreover, cytophilic antibodies with anti-tumour specificity will act as 'unblocking' antibodies. These predictions are at

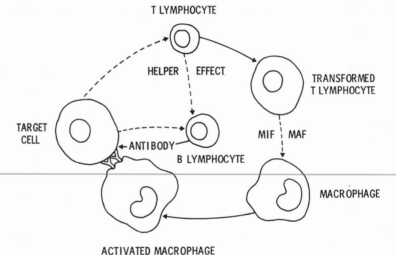

FIG. 4. Interactions of lymphocytes and macrophages in antitumour immunity. Antigens on the surface of the target cell stimulate transformation of T lymphocytes able to recognize them. During the transformation macrophage migration inhibitory factor (MIF) and macrophage activation factor (MAF) are liberated. The latter stimulates hydrolase synthesis, and the macrophage becomes activated. Concurrently, B lymphocytes release antitumour antibody which can become attached to the target cell and then react (through Fc) with the receptors on the macrophage plasma membrane. Alternatively the antibodies could be cytophilic, attached to the macrophage first and reacting with the target cell (through Fab) second. In either case intimate contact between activated macrophage and target cell is achieved and the target cell is killed.

present being tested in my laboratory, and preliminary results look promising. The overall interpretation is summarized in Fig. 4.

APPLICATION TO PARASITES

The question can reasonably be asked what all this has to do with immunity against helminths and other parasites—apart from casual observations on the curious effects of parasitization on tumour growth. Perhaps an adequate answer will be the affirmation of my belief that immunity against metazoan parasites will prove to be due to mechanisms similar to those operating against tumours and allografts. In particular, close attention should be paid to the role of activated macrophages and antibodies acting in concert.

The enhancing effect of serum from rats immune to *Nippostrongylus* on tumour growth in rats would be consistent with the view that relatively high levels of antiparasite cytophilic antibodies are formed. This should be looked for, and the possible role of such antibodies examined. We have presented evidence for a role of T cells in protection against malaria (Brown *et al.*, 1968) and malarial antigen in Freund's complete adjuvant confers more complete protection against the parasites than does antigen in incomplete adjuvant (Brown, 1969). These observations could imply that activated macrophages, acting synergistically with the right kind of antibody, are better able to kill the parasites than in the absence of either component. There is no need to cite further possible examples: each one must be worked out in detail to reveal the underlying mechanism; but the information gathered on immunity against tumours and allografts may well come to have application in the study of anti-parasite immunity.

REFERENCES

ALEXANDER, P. and EVANS, R. (1971). Endotoxin and double-stranded RNA render macrophages cytotoxic. *Nature New Biology* **232**: 76

ALLISON, A. C. (1971). New cell antigens induced by viruses and their biological significance. *Proceedings of the Royal Society, Series B* **177**: 23

ALLISON, A. C., DAVIES, P. and DE PETRIS, S. (1971). The role of contractile microfilaments in macrophage movement and endocytosis. *Nature New Biology* **232**: 153

AMOS, D. B. (1962). Host responses to ascites tumours. *2nd International Symposium on Immunopathology*, p. 210. Basel: Schwabe

BANSAL, S. C. and SJÖGREN, H. O. (1971). 'Unblocking' serum activity *in vitro* in the polyoma system may correlate with antitumour effects of antiserum *in vivo*. *Nature New Biology* **233**: 76

BENNETT, B. (1965). Specific suppression of tumor growth by isolated peritoneal macrophages from immunized mice. *Journal of Immunology* **95**: 656

BERKEN, A. and BENACERRAF, B. (1966). Properties of antibodies cytophilic for macrophages. *Journal of Experimental Medicine* **123**: 119

BOSMANN, H. B., HAGOPIAN, A. and EYLAR, E. H. (1969). Cellular membranes: the biosynthesis of glycoprotein and glycolipid in HeLa cell membranes. *Archives of Biochemistry and Biophysics* **130**: 573

BOYDEN, S. V. (1964). Cytophilic antibody in guinea-pigs with delayed-type hypersensitivity. *Immunology* **7**: 474

BOYSE, E. A., OLD, L. J. and OETTGEN, H. F. (1969). Tumour immunology. In *Textbook of Immunopathology*, Vol. II, p. 768. P. A. Miescher and H. J. Muller-Eberhard (eds). New York: Grune & Stratton

BROWN, I. N., ALLISON, A. C. and TAYLOR, R. B. (1968). *Plasmodium berghei* infections in thymectomized rats. *Nature, London* **219**: 292

BROWN, K. N. (1969). Immunity to protozoal infections. *Proceedings of the Royal Society of Medicine* **62**: 301

CEROTTINI, J. C., NORDIN, A. A. and BRUNNER, K. T. (1970). Specific *in vitro* cytotoxicity of thymus-derived lymphocytes sensitized to alloantigens. *Nature, London* **228**: 1308

COHN, Z. A. and PARKS, E. (1967). The regulation of pinocytosis in mouse macrophages. II. Factors inducing vesicle formation. *Journal of Experimental Medicine* **125**: 213

EBASHI, S. and ENDO, M. (1968). Calcium ion and muscle contraction. *Progress in Biophysics and Molecular Biology* **18**: 123

EVANS, R. and ALEXANDER, P. (1970). Co-operation of immune lymphoid cells with macrophages in tumour immunity. *Nature, London* **228**: 620

GORER, P. A. (1956). Some recent work on tumour immunity. *Advances in Cancer Research* **4**: 149

GRANGER, G.A. and WEISER, R. S. (1964). Homograft target cells: specific destruction *in vitro* by contact interaction with immune macrophages. *Science* **145**: 1427

GRANGER, G. A. and WEISER, R. S. (1966). Homograft target cells: contact destruction *in vitro* by immune macrophages. *Science* **151**: 97

GREEN, H., BARROW, P. and GOLDBERG, G. (1959). Effect of antibody and complement on permeability control in ascites tumor cells and erythrocytes. *Journal of Experimental, Medicine* **110**: 699

HARDING, B., PUDIFIN, D. J., GOTCH, F. and MACLENNAN, I. C. M. (1971). Cytotoxic lymphocytes from rats depleted of thymus processed cells. *Nature New Biology* **232**: 80

HELLSTRÖM, I. and HELLSTRÖM, K. E. (1970). Colony inhibition studies on blocking and non-blocking serum effects on cellular immunity to Moloney sarcomas. *International Journal of Cancer* **5**: 195

HELLSTRÖM, I., HELLSTRÖM, K. E., PIERCE, G. E. and FEFER, A. (1969). Studies on immunity to autochthonous mouse tumors. *Transplantation Proceedings* **1**: 90

HELLSTRÖM, K. E. and HELLSTRÖM, I. (1969). Cellular immunity against tumor antigens. *Advances in Cancer Research* **12**: 167

HESS, M. W. and LUSCHER, E. F. (1970) Macrophage receptors for IgG aggregates. *Experimental Cell Research* **59**: 193

HIBBS, J. B., LAMBERT, L. H. and REMINGTON, J. S. (1971). Tumour resistance conferred by intracellular protozoa. *Journal of Clinical Investigation* **50**: 45a (abstracts)

HIBBS, J. B., LAMBERT, L. H. and REMINGTON, J. S. (1972). Macrophage mediated non-specific cytotoxicity—possible role in tumour resistance. *Nature, New Biology* **235**: 48

HUBER, H., DOUGLAS, S. D., NUSBACHER, J., KOCHWA, S. and ROSENFIELD, R. E. (1971). IgG subclass specificity of human monocyte receptor sites. *Nature, London* **229**: 419

HUMPHREY, J. H. and DOURMASHKIN, R. B. (1969). The lesions in cell membranes caused by complement. *Advances in Immunology* **11**: 75

KELLER, R. and JONES, V. E. (1971). Role of activated macrophages and antibody in inhibition and enhancement of tumour growth in rats. *Lancet*, **ii**: 847

KELLER, R., OGILVIE, B.M. and SIMPSON, E. (1971). Tumour growth in nematode-infected animals. *Lancet* **i**: 678

LAW, L. W. (1966). Studies of thymic function with emphasis on the role of the thymus in oncogenesis. *Cancer Research* **26**: 551

LAY, W. H. and NUSSENZWEIG, V. (1968). Receptors for complement on leucocytes. *Journal of Experimental Medicine* **128**: 991

MACKANESS, G. B. (1970). Cellular immunity. In *Mononuclear Phagocytes*, p. 461. R. van Furth (ed.). Oxford: Blackwell

NELSON, D. S. (1969). *Macrophages and Immunity*. Amsterdam: North-Holland Press

ORR, T. S. C. and BLAIR, A. M. J. N. (1969). Potentiated reagin response to egg albumin and conalbumin in *Nippostrongylus brasilensis* infected rats. *Life Sciences* **8**: 1073

RABINOWITCH, M. (1967). The dissociation of the attachment and ingestion phases of phagocytosis by macrophages. *Experimental Cell Research* **46**: 19

RUBIN, R. P. (1970). The role of calcium in the release of neurotransmitter substances and hormones. *Pharmacological Reviews* **22**: 389

SORKIN, E. and BOYDEN, S. V. (1959). Studies on the fate of antigen *in vitro*. I. The effect of specific antibody on the fate of I^{131} trace labelled human serum albumin *in vitro* in the presence of guinea pig monocytes. *Journal of Immunology* **82**: 332

WEATHERLY, N. F. (1970). Increased survival of Swiss mice given sublethal infections of *Trichinella spiralis*. *Journal of Parasitology* **56**: 748

YASHPHE, D. J. (1971). Immunological factors in non-specific stimulation of host resistance to syngeneic tumors. *Israel Journal of Medical Science* **7**: 90

AUTHOR INDEX

Numbers in heavy type refer to the pages on which references are listed at the end of each paper.

109

SUBJECT INDEX

Adenosine triphosphate, action on parasites, 83–84
Adhesion, between organisms, factors controlling, 1–21
  breaking, 2–3
  cells, collision, 2
    control systems, 10–13
    enzymic action, 12–13
    localisation, 17–18
    non-coalescence, 16–17
    specificity, 13–18
      quantitative method for testing, 15–16
    'temporal', 15
  colloid, DVLO theory, 6–10
  formation, 3
  of red blood cells, 11–12
  synthesis, 8–9
Agglutination, merozoites, 83
Antibodies, blocking and unblocking, 99
  cytophilic, 97–98, 102–103
  and lysis of cells, 94–95
  reaginic, 101
Antigens, 76
  exoantigen, 73
  host, see Host antigens
  variant, Trypanosomes, 73, 74

Basal membrane complex, of cestode tegument, 47–48
Brownian motion, and formation of adhesions, 2–3

Caryophyllaeus laticeps, vesicular inclusions, 46
Cells, red blood, effect on adhesion, 11–12
  involvement as host antigens, 33–35

surface, nature, 3
  see also Adhesion, and headings throughout
Cercariae, schistosome, see Schistosome
Cestodes, digestive-absorptive surface changes, 41–70
  tegument as digestive-absorption surface, 58–66
  see also Tegument, Membrane and headings throughout
Collision of cells, and formation of adhesion, 2
Collision efficiency, 15–16
Complement, and lysis of cells, 94–95
Cryptosporidium, 81, 82
Cytoplasmic inclusions, of cestode tegument, 46

Desmosomes, 77–78
Digestion, extracellular, of cestode tegument, 58
  intracellular, of cestode tegument, 58
  membrane, and acquisition of 'host-like' antigens, 66–67
    of cestode tegument, 59–60
    in other cestodes, 64–66
    in E. granulosus and other cestodes, 60–66
Digestive-absorptive surface, cestodes, changes, 41–70
  cestode tegument, 58–66
Diphyllobothrium differentiation, 49–51
  plerocercoid/adult transformation, 50–51

proceroid / plerocercoid transformation, 49–50
  vesicular inclusions, 46
Diphyllobothrium erinacei, microtriches, 45
Diphyllobothrium latum, external plasma membrane, 44
Diplogonoporus grandis, vesicular inclusions, 46
Dipylidium caninum, pore canals, 47
  vesicular inclusions 46
DLVO colloid adhesion, 6–10

Echinococcus granulosus, adult/larval differentiation, 51–58
  basal membrane complex, 48
  contact stimulus, 52–56
  limiting membrane, 44–45
  membrane digestion, 60–64
  protoscolex/adult differentiation, ultrastructural changes, 56–58
  surface membrane involvement in induction of strobilisation, 51–52
  vesicular inclusions, 46
Ectoparasitic protozoa, see Protozoa, ectoparasitic
Endocytosis, requirement for, 96–97
Entamoeba histolytica, surface coat, 72–73
Enzymes, membrane-bound, in E. granulosus, 60–64
Exoantigen, 73
Exocytosis, requirements, 96–97

112